Setting the Level and Annual Adjustment of Military Pay

BETH J. ASCH, MICHAEL G. MATTOCK, TROY D. SMITH, JASON M. WARD

Prepared for the 13th Quadrennial Review of Military Compensation
Approved for public release; distribution unlimited

RAND NATIONAL DEFENSE RESEARCH INSTITUTE

For more information on this publication, visit www.rand.org/t/RRA368-1

Library of Congress Cataloging-in-Publication Data is available for this publication.
ISBN: 978-1-9774-0585-2

Published by the RAND Corporation, Santa Monica, Calif.
© Copyright 2020 RAND Corporation
RAND® is a registered trademark.

Cover: guvendemir/Getty Images

Support RAND
Make a tax-deductible charitable contribution at
www.rand.org/giving/contribute

www.rand.org

Preface

Federal law mandates that every four years the Secretary of Defense conduct an assessment of the military compensation system, resulting in a quadrennial review. The 13th such review, the Thirteenth Quadrennial Review of Military Compensation (13th QRMC), began in 2018 and is focusing on several topics, including an assessment of a salary system to replace the current basic pay and allowance system and an assessment of a time-in-grade pay table to replace the current time-in-service pay table to increase incentives for performance. Two other topics being considered by the 13th QRMC are the continued relevance of the 70th percentile of civilian earnings, which is currently being used as a benchmark for setting military pay, and the continued use of the Employment Cost Index (ECI) to guide the annual military pay adjustment. Analysis of military and civilian pay in the 1990s indicated that pay at around the 70th percentile had historically been necessary to enable the military to recruit and retain the quality and quantity of personnel required. But more recent research has found that military pay has exceeded the 70th percentile benchmark in recent years, raising the question of the continued relevance of this benchmark. Research from the early 1990s suggested that an alternative to the ECI, the Defense Employment Cost Index (DECI), would be more relevant to military personnel. The 7th QRMC, in 1992, did not recommend the use of this alternative but did recommend that the U.S. Department of Defense continue the development of the DECI. That development did not occur.

The 13th QRMC asked the RAND Corporation to provide input on the setting of the level of military pay, the relevance of the 70th percentile, and the use of the DECI on the annual adjustment to military pay. This report describes the results of these analyses. It should be of interest to those concerned about the setting of military pay and the annual adjustment. The research reported here was completed in June 2020 and underwent security review with the sponsor and the Defense Office of Prepublication and Security Review before public release.

The research was sponsored by the 13th QRMC and conducted within the Forces and Resources Policy Center of the RAND National Security Research Division (NSRD), which operates the National Defense Research Institute (NDRI), a federally funded research and development center sponsored by the Office of the Secretary of Defense, the Joint Staff, the Unified Combatant Commands, the Navy, the Marine Corps, the defense agencies, and the defense intelligence enterprise.

For more information on the RAND Forces and Resources Policy Center, see http://www.rand.org/nsrd/frp or contact the center director (contact information is provided on the webpage).

Contents

Figures

Tables

Summary

A key objective of the military compensation system is to enable the military to attract and retain personnel in sufficient numbers and quality to meet its manpower requirements, and to do so efficiently. With respect to the quality of military enlisted recruits, the U.S. Department of Defense's (DoD's) stated requirement is that 90 percent of recruits each year and in each service must be at least high school graduates and that 60 percent must be above average in terms of aptitude. To meet these requirements, past commissions, including the work of the Ninth Quadrennial Review of Military Compensation (9th QRMC) in 2002, found that military pay should be set at around the 70th percentile of the earnings for similar civilians. The 9th QRMC found that pay must be higher than average civilian pay because of the unusual demands and arduous nature of military service. The specific benchmark of the 70th percentile was chosen based on research from the 1990s that used data from the late 1980s and mid-1990s. Setting military pay involves not just setting the level but also determining the annual military pay adjustment. By law, the annual increase in military basic pay is guided by changes in the Employment Cost Index (ECI), a measure of the growth in private-sector employment costs computed by the Bureau of Labor Statistics within the U.S. Department of Labor. The ECI is also used to guide changes in the pay of federal General Schedule civil service employees, and, since 1967, the annual pay adjustment for military pay has been tied to the adjustment for General Schedule employees.

More recent research, including the work of the 11th QRMC in 2012, has found that military pay, as measured by regular military compensation (RMC), far exceeds the 70th percentile and has done so for quite some time.[1] This finding leads to the question of whether the level of military pay is being set efficiently, and, in particular, whether RMC is too high relative to the earnings of similar civilians. Or, put differently, it raises the question of whether the 70th percentile is still relevant as the benchmark, and, if not, what the new benchmark for setting the level of military pay should be.

Regarding the ECI, research from the early 1990s (Hosek et al., 1992) found that the ECI does not accurately measure the opportunity wages of active duty personnel or perform well in terms of tracking recruiting and retention outcomes. The use of the ECI led to a "pay paradox" in the 1990s, when growth in basic pay fell well short of changes in the ECI, even though the services were meeting their recruiting and retention objectives. The research recommended the use of an alternative index, the Defense Employment Cost Index (DECI), that better reflected the demographics of military personnel and so more accurately measured the opportunity wages of active duty personnel. Not surprisingly, the DECI performed better than the ECI in

---

[1] As we describe in Chapter One, RMC includes military basic pay, the basic allowance for housing, the basic allowance for subsistence, and the tax advantage associated with the tax-free status of allowances.

terms of tracking recruiting and retention outcomes. The 7th QRMC in 1992 chose not to recommend that the DECI replace the ECI but recognized its advantages and recommended that its continued development be supported by DoD. No such development has occurred since then.

The 13th QRMC requested that the RAND National Defense Research Institute assess the continued relevance of the 70th percentile as a benchmark for setting the level of military pay and the advantages and disadvantages of the DECI. It also requested that RAND provide computer code for computing the RMC percentile and for computing the DECI, for DoD's future use. This report summarizes our analysis and findings and includes, in appendixes, the relevant computer code. Our approach involved using military personnel and pay data provided by the Defense Manpower Data Center (DMDC), active duty personnel survey data provided by DoD's Office of People Analytics, and Current Population Survey data from the Bureau of Labor Statistics. Our analysis builds on and extends previous studies that have computed RMC percentiles (Hosek, Asch, and Mattock, 2012; Hosek et al., 2018; Smith, Asch, and Mattock, 2020) and the Hosek, Asch, and Mattock (2012) study that computed the DECI and compared it with the ECI.

RMC Percentiles: Findings, Implications, and Recommendations

We computed weighted RMC percentiles for enlisted personnel and officers, adjusting for the education distribution of military personnel. The RMC percentile today exceeds the 70th percentile benchmark set based on data from the late 1980s and mid-1990s, reflecting the relatively fast military pay growth from the late 1990s to 2010, as well as a downward trend in real civilian wages. The first two rows of Table S.1 show the weighted average percentiles for enlisted personnel in 2018 and in two benchmark periods, 1988–1989 and 1993–1997. We use two data sources to compute weights: administrative personnel records from DMDC's Active Duty Master Files (ADMF) and survey responses from DoD Status of Forces Surveys (SOFS). The former data have the advantage of a long time series, permitting computation of percentiles back to the 1980s, and the latter data more accurately reflect the education distribution of military personnel. The third and fourth rows of Table S.1 show similar weighted average percentiles for officers.

The weighted average RMC percentiles for enlisted personnel were 88.8 and 85 percent in the most recent data period, 2018–2019, using, respectively, the ADMF versus the SOFS for computing earnings weights. Regardless of the data source for computing weights, these earnings percentiles exceed the 70th percentile benchmark. They also exceed the 75.2 average enlisted percentile for the 1993–1997 period and the 73.9 percentile for 1988–1989 period that we estimate. However, we use a different methodology from the past studies that were the basis for setting the 70th percentile benchmark. Similarly, the weighted average of the RMC percentiles for officers were 78 and 77, respectively, using the different data sources. These also exceed the average officer RMC percentile for the 1993–1997 and 1988–1989 periods that we estimate, equal to 71 for both periods.[2]

---

[2]   The RMC percentiles are computed based on observed educational achievement, as measured by the ADMF or by the SOFS. These computations do not provide an assessment of the educational attainment that is required for military duties. To the extent that military personnel seek additional education to improve their post-service military earnings rather than to meet requirements for conducting their military duties, the observed educational attainment of personnel could

Table S.1
Weighted Average Enlisted and Officer RMC Percentiles, and DoD Recruiting and Retention Outcomes, 2018–2019 and Selected Benchmark Years

	Average 2018–2019	Average 1993–1997	Average 1988–1989
All Services			
Enlisted RMC percentile (ADMF weights)a	88	75	73
Enlisted RMC percentile (SOFS weights)a	85		
Officer RMC percentile (ADMF weights)[a]	78	71	71
Officer RMC percentile (SOFS weights)[a]	77		
Recruiting			
Percent Tier 1	96.9	95.4	92.5
Percentage AFQT Category I–IIIA	69.0	70.4	65.5
Retention			
Enlisted Continuation Rate at YOS 4	73%	63%	62%
Enlisted Continuation Rate at YOS 8	85%	84%	87%
Officer Continuation Rate at YOS 8	91%	89%	88%
Resources			
Recruiters	14,367	11,967	12,796
Enlistment Bonuses[b]	$530,191	$48,357	$113,369
Reenlistment Bonuses[b]	$1,092,816	$386,126	$865,952
All Services Except Army			
Recruiting			
Percentage Tier 1	98.6	96.2	93.5
Percentage AFQT Category I–IIIA	75.0	70.9	66.9
Retention			
Enlisted Continuation Rate at YOS 4	76%	58%	64%
Enlisted Continuation Rate at YOS 8	85%	86%	87%
Officer Continuation Rate at YOS 8	93%	87%	88%
Resources			
Recruiters	6,609	7,047	7,229
Enlistment Bonuses[b]	$109,584	$19,903	$40,197
Reenlistment Bonuses[b]	$639,936	$312,807	$673,932

SOURCE: Office of the Under Secretary of Defense for Personnel and Readiness and tabulations provided by DMDC.
NOTES: AFQT = Armed Forces Qualifying Test; YOS = years of service.
[a] RMC percentile for 2018–2019 is for 2018 only.
[b] Constant 2019 dollars. Enlistment bonus totals exclude Air Force.

exceed the military's educational requirements. If so, the RMC percentiles reported in Table S.1 will be understated. That is, if DoD educational requirements are lower than the observed educational attainment, the true RMC percentiles would be even higher than those reported in Table S.1.

xvi   Setting the Level and Annual Adjustment of Military Pay

A key question is whether higher RMC percentiles are needed today to meet DoD's stated recruiting requirements, or at least to achieve the outcomes observed in the 1980s and 1990s, the period relevant to the decision to set the 70th percentile as a benchmark. To address this question, we consider recruit quality and retention outcomes. Table S.1 shows statistics on the quality of accessions and average retention rates of enlisted personnel at 4 and 8 years of service (YOS) and of officers at 8 YOS. The table shows results for DoD overall (top panel) and DoD excluding the Army (bottom panel). Army recruit quality has followed a different pattern than the rest of the services, so results differ if the Army is excluded.

Comparing the most recent outcomes for which we have data, 2018 and 2019, with those for 1988–1989 and 1993–1997—the periods that were relevant to the studies that led to the 70th percentile being set as the benchmark for military pay levels—we find that, relative to the earlier periods, recruit quality across DoD, measured in terms of the percentage of accessions that are AFQT categories I–IIIA and are Tier 1, is about the same in recent years. However, when the Army is excluded from the measurement, recruit quality is higher in recent years. Furthermore, retention, measured in terms of the continuation rate at YOS 4 for enlisted personnel and at YOS 8 for both enlisted and officers is also higher, regardless of whether the Army is excluded. As the other services increased recruit quality in more recent years, the Army kept quality close to the DoD benchmarks, for reasons that are unknown. The higher quality of recruits and the higher continuation rates are notable because the adult unemployment rate was lower in 2018–2019 than in the earlier periods, implying that recruiting and retention likely were more difficult for the services to sustain. Notably, recruit quality exceeded the DoD benchmarks both more recently and in the earlier periods.

The higher recruit quality and retention could also reflect increases in recruiting and retention resources that also affect outcomes, such as enlistment bonuses, reenlistment bonuses, and recruiters. Past studies have found that outcomes are positively related to these resources. Enlistment bonuses increased in more recent years relative to the earlier periods, in real dollars, but reenlistment bonuses were about the same in real dollars as in the 1988–1989 period, and the number of recruiters was relatively unchanged, when the Army is excluded from the tabulation.

Implications

These findings suggest that the RMC percentiles may be too high, since recruit quality today exceeds DoD's benchmarks and, further, quality and retention both exceed the levels observed during the late 1980s and mid-1990s, when the 70th percentile was established. That said, these findings do not necessarily imply that the 70th percentile continues to be the appropriate benchmark. In other words, it does not necessarily follow that the growth of military pay should be slowed in the future to the point that the 70th percentile benchmark is achieved, for two reasons. First, changes in defense threats, readiness requirements, and military technology may have shifted manpower requirements toward higher-aptitude recruits in some services. If this is the case, then DoD should increase the recruit quality benchmarks and consider resources other than higher pay, such as bonuses and special and incentive pays, to achieve those benchmarks. Second, even if current recruit quality benchmarks remain valid, there are reasons to believe that the recruiting environment is more difficult than it was in earlier periods, due to factors that are not transitory (such as a historically low unemployment rate), making recruiting requirements more difficult to achieve today than in earlier years. The literature, including the March 2020 report of the National Commission on Military, National, and Public Service, has expressed concern about the so-called "military-civilian divide" and its adverse effects on military recruiting. In addition, the recruiting environment may be more difficult because few American youth (less than a third) would meet enlistment standards without a waiver, and some of the factors that lead to disqualification are increasing in prevalence in the civilian youth market. For example, rising youth obesity is particularly concerning.

Recommendations

Together, these factors suggest that an RMC percentile benchmark above the 70th percentile is appropriate. But if the 85th percentile for enlisted personnel and the 77th percentile for officers are too high (as measured using the SOFS weights), and the 70th is too low, given the benchmarks for outcomes, what is the right number? We do not have a specific number to offer, but we believe a figure of around the 75th to 80th percentile for enlisted personnel is likely to meet existing recruit quality objectives. Similarly, we believe that a figure of around the 75th percentile for officers is likely to be appropriate. Given the evidence that increasing military pay is associated with improved recruit quality and retention, increasing the benchmark would address a more challenging environment and help ensure that the services can continue to meet and even exceed DoD's objectives for recruit quality and higher retention rates.

DoD should continue to monitor the RMC percentile, along with recruit quality, retention, and enlisted and officer retention, to ensure that the RMC percentile is at or around the new benchmark. We recommend that, in computing the RMC percentile, DoD use the SOFS to compute weights, because these data most likely provide more accurate information on members' educational attainment. We provide computer code in the appendixes for DoD to use in computing the RMC percentile in the future.

ECI Versus the DECI: Findings, Implications, and Recommendations

Building on the 1992 RAND report *A Civilian Wage Index for Defense Manpower*, by Hosek et al. (1992) (henceforth "Hosek 1992"), we extended and updated the DECI and compared it with the ECI and an index of the growth in basic pay or BPI (basic pay index). Figure S.1 plots the time series of the BPI, the ECI, and the DECI from the baseline year of 1982. The figure shows evidence of a negative pay gap between basic pay and the ECI (basic pay growth lagged behind the ECI) that began immediately after the baseline year of 1982 and that did not close until the above-ECI increases in basic pay between 2000 and 2010. In contrast, two decades after the baseline year—until around 2000—we find no systematic evidence of a pay gap using the DECI. On the other hand, by 2008, at the onset of the Great Recession, the growth in basic pay had created a large positive gap between the BPI and the DECI, reflecting slower pay growth relative to military personnel for civilians with demographics similar to military personnel. During the period of the Great Recession and its aftermath, the positive gap between the BPI and DECI grew even larger, reflecting stagnation in earnings growth for civilians with similar demographics to military personnel. At its largest, in 2011, the value of the positive gap between the BPI and the DECI was greater than 25 percent of the overall index value.[3]

---

3 As noted in footnote 2, the observed educational attainment of military personnel could exceed the military's educational requirements if individuals attain greater education to improve their post-service earnings rather than to meet

Figure S.1
Comparing ECI- and DECI-Based Civilian Earnings Growth over Time, Using 1982 as the Baseline Year

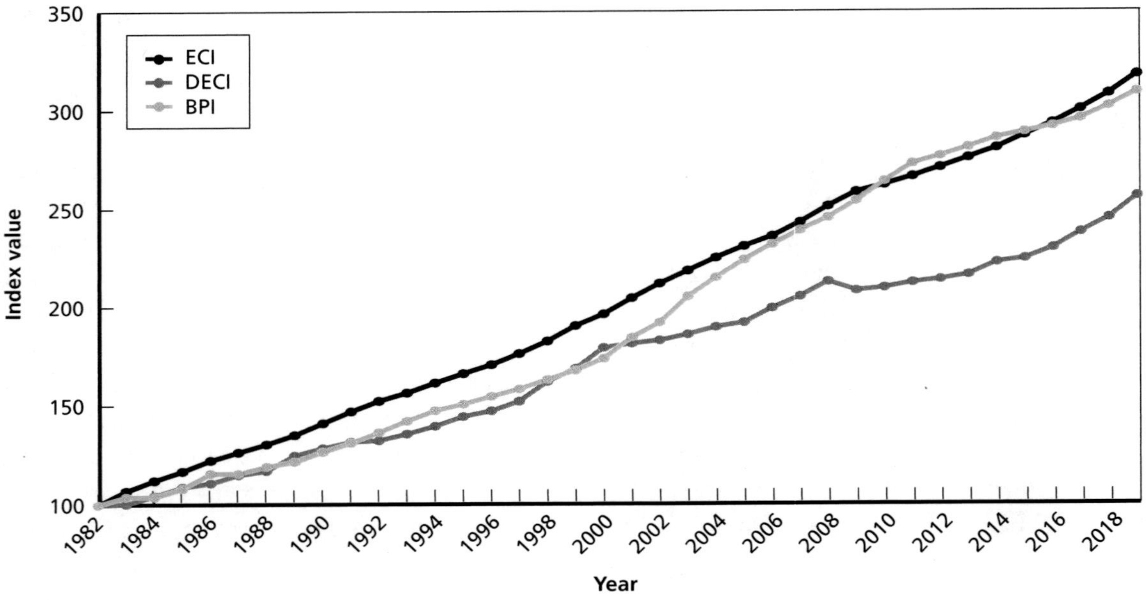

SOURCES: Current Population Survey (CPS) Outgoing Rotation Group (ORG) data (April–September, 1982–2019) from the Integrated Public Use Microdata Series (IPUMS (Flood et al., 2020); ECI data from Bureau of Labor Statistics; BPI data from OUSD(P&R) (2018).
NOTES: DECI cells comprise eight age groups and four education groups. Military weights are generated from ADMF data aggregating officers and enlisted from four services but omitting Navy officers because of poor coding of education.

We note that the size of differences in the DECI versus the ECI and the BPI, and the resulting guidance the DECI provides for setting the annual pay adjustment, is affected by the choice of the base year—1982 in Figure S.1. The importance of this aspect of the analysis is difficult to overstate. Indices, by their nature, can only provide guidance on rates of change, and have nothing to say regarding the comparability of earnings in levels. The "correct" baseline year is a policy choice that must reflect the views of decisionmakers about when the relative level of military pay to civilian earnings was set appropriately. We chose the year 1982 because it was a year when military pay was viewed as competitive with civilian pay because it followed two large pay changes that occurred in 1981 and 1982 and that were intended to restore competitiveness.

The DECI has several advantages: the use of underlying data that reflect labor market outcomes from an employee, rather than an employer, perspective; weighting that reflects the age, educational attainment, and gender composition of military personnel; and the flexibility to assess paths of earnings growth for specific subgroups of interest within the overall active

cational requirements if individuals attain greater education to improve their post-service earnings rather than to meet military requirements for conducting their duties. If so, the DECI computation will give greater weight to better-educated civilians than what would occur if we had incorporated DoD's educational requirements. Because better-educated civilians have experienced faster pay growth, the DECI shown in Figure S.1 is higher than what we would have computed if had we used DoD's educational requirements. If the DECI is used to guide the annual pay adjustment, it would imply faster military pay growth than what would have been suggested by a DECI based on lower educational attainment.

duty force. These advantages lend face validity to the DECI for the purpose of adjusting basic pay. But another advantage of the DECI is that it outperforms the ECI in terms of tracking personnel outcomes. For example, we find that the DECI explains more of the variation in military personnel outcomes than does the ECI. For example, Table S.2 shows the R^2 of regressions of the percentage of accessions that are high-quality on the ratio of BPI to DECI versus the ECI. The explanatory power of the DECI as measured by the regressions' R^2 is two to five times greater than that of the ECI, especially when we exclude the Army from the analysis. For example, excluding the Army, the R^2 is 0.830 for using the DECI versus 0.365 using the ECI. The higher explanatory power of the DECI relative to the ECI lends additional validity to the DECI.

That said, the DECI was rejected by the 7th QRMC and DoD because of concerns about its accuracy, timeliness, cost, flexibility, and relevance. We considered these issues in light of the availability of data beyond the early 1990s and advances in computer technology that make processing these data easier. We find that most of the critiques of the DECI have been addressed by advances in data availability and computing power in recent decades.

With respect to accuracy, the Current Population Survey (CPS) data that underlie the DECI come from the Bureau of Labor Statistics, the same source as the National Compensation Survey data that underlie the ECI. Furthermore, CPS data are the source of national measures of unemployment and labor force participation. One issue regarding accuracy of data is the accuracy of the military data used to weight the CPS earnings data. The ADMF may have some data quality issues with respect to members' educational achievement, and the SOFS may be a viable alternative. But, notably, we find that the relative trends in the DECI, ECI, and BPI are quite similar regardless of whether the ADMF or SOFS is used. Regarding timeliness, at the time that the Hosek 1992 study was published, data access was a significant hurdle for all kinds of quantitative analyses, even those that used data from a ubiquitous survey such as the CPS. Nearly three decades later, the issue of timely access to data such as the CPS's has been resolved by advances in computing power and connectivity. Furthermore, unlike the

Table S.2
R^2 for Regressions Between High-Quality Accession Rates and Enlisted DECI Versus ECI

	R^2
Percentage High-Quality Accessions—All Services	
BPI to DECI ratio	0.493
BPI to ECI ratio	0.107
Percentage High-Quality Accessions—All Services Except the Army	
BPI to DECI ratio	0.830
BPI to ECI ratio	0.365

NOTES: Estimates are from a bivariate regression of high-quality accession rate (in percent and as defined in each panel) on the indicated ratio (DECI or ECI). Accession data are from the Office of the Under Secretary of Defense. The DECI measure uses weighting based on the composition of enlisted service members. See Chapter Four for further details on data sources and index construction.

Hosek 1992 study, which used the March CPS supplement for data on civilian earnings, we use monthly CPS data collected through the end of September to generate a DECI (or multiple DECIs) by late October. This advance in data availability eliminates any meaningful difference in the timeliness of the DECI relative to the ECI. Regarding cost, the ECI has the advantage of being generated using the time, resources, and expertise of another federal agency. However, generating the DECI is readily within the grasp of the Office of the Secretary of Defense. For example, it could be computed by either the Office of People Analytics within the Office of the Under Secretary of Defense for Personnel and Readiness (OUSD/P&R) or by the Office of Cost Assessment and Program Evaluation within DoD. As the group that provides independent analytics for the Secretary of Defense, computation of the DECI by the latter group could help ensure that the computation of the DECI is performed independently of OUSD/P&R. The software code written to generate the DECIs for this study is provided in Appendix E of this report. This software could further be used to assist with a periodic audit of the DECI to ensure its continued accuracy.

Regarding relevance, because the DECI uses weights that reflect the demographics of military personnel, it captures civilian earnings growth more applicable to military personnel. Furthermore, the changes in civilian earnings measured by the DECI impart more relevant information with respect to accession and retention outcomes. As discussed above, we find that the DECI outperforms the ECI in terms of tracking personnel outcomes. Particularly relevant to the period examined by the Hosek 1992 study, we find that the paradox of the "pay gap" of the 1980s and 1990s, when civilian earnings were measured by the ECI, essentially disappeared when the DECI was used in its place. The DECI is also more relevant to the setting of the pay adjustment for military personnel because it is more sensitive to the effect of macroeconomic fluctuations on less experienced, lower-wage workers than the ECI, as the ECI averages over all workers, young and old. A lack of sensitivity to fluctuations in civilian earnings based on macroeconomic shocks was portrayed by the 7th QRMC as a virtue of the ECI and a potential drawback of the DECI. But we find that the ECI tends to mute, reduce, or even negate civilian pay growth during recessions, thereby muting the effect of these changes in average civilian earnings on the suggested annual military pay adjustment.[4] Using the ECI helps to limit any negative impact of recessions on the pay raise for military personnel, but using an index that is insensitive to significant changes in the civilian labor market to set the annual pay raise is not a cost-effective policy from the standpoint of the taxpayer. During recessions, recruiting and retention tend to improve, implying that pay raises should be more limited during recessions than the ECI might suggest. By being more sensitive to macroeconomic fluctuations, the DECI would provide more accurate and relevant guidance for sustaining recruiting and retention at lower cost.

Finally, the DECI is more flexible because it can be computed for subgroups of interest, such as for enlisted personnel and officers or for personnel in the cyber community. This is in contrast to the ECI, which is employer-based and can only trace broad industry subgroups,

---

[4]   One way to empirically measure these differences in sensitivity to recessions is to estimate a regression model that regresses the annual percentage change in the relevant index (the ECI or the DECI) on the percentage change in the unemployment rate. The results of these estimates show that the DECI is highly correlated with the unemployment rate (a 1 percent increase in the unemployment rate is correlated with a 1.33 percent decrease in the DECI with an estimated standard error of 0.35, while the same coefficient using the percent change in the ECI is 0.161 with an estimated standard error of 0.20).

rather than employee subgroups. Additionally, the ECI is not produced for the military, so it cannot be customized to fit any specific needs related to setting military compensation.

Recommendations

The DECI is a promising alternative, or supplement, to the ECI because of the advantages of the DECI relative to the ECI, because the DECI has continued to perform well empirically in explaining key military manpower outcomes, and because the disadvantages of the DECI noted in the past are less relevant today. We recommend that DoD compute the DECI each year, either as a supplement to the ECI or as a replacement, to impart information about civilian wage growth that is relevant to military personnel.

Wrap Up

Any assessment about the adequacy of the level of military pay and the appropriate adjustment of military pay should incorporate information about whether the military is meeting its manpower objectives and the state of military recruiting and retention outcomes. It should also incorporate information about whether changes in outcomes reflect transitory factors, such as changes in the unemployment rate, are specific to a relatively small community of military personnel and can be addressed through bonuses and special and incentive pays or changes in personnel policy, or are permanent and widespread factors that are best addressed through changes in the level of military pay.

That said, decisions about the level and annual adjustment need a starting point, and the starting point for decades has been the 70th percentile of earnings of similar civilians. Furthermore, the starting point for the annual pay adjustment has been the change in the ECI, an index that does not reflect the growth of earnings of similar civilians.

Our analysis suggests that these starting points can be improved upon so that military pay can be set more efficiently. The RMC percentile should increase above the 70th percentile, and we believe that it should be somewhere between the 75th and 80th percentile of civilian earnings and that DoD should either supplement the ECI or replace the ECI with the DECI. Due to significant changes in both data availability and computing power, the DECI as well as the RMC percentile can now be generated in a timely fashion using resources and expertise already available within DoD.

Acknowledgments

We are indebted to Thomas Emswiler and Colonel Brunilda Garcia, director and associate director, respectively, of the 13th QRMC. We are also grateful to Jerilyn Busch, director of the Office of Compensation within the office of the Under Secretary of Defense for Personnel and Readiness, and to Don Svendsen also within that office, for their assistance to this project including their help in accessing historical, hard-copy, information on military pay. We would also like to thank Christopher Arendt and Dennis Drogo in the Office of Accession Policy, as well as Michael DiNicolantonio in the Office of People Analytics, both within the office of the Under Secretary of Defense for Personnel and Readiness. They provided requested tabulations and data, including tabulations from the Status of Forces Survey. At RAND, we wish to thank Craig Bond for comments on an earlier draft and Alice Shih, Christine DeMartini, Tony Lawrence, and Jonas Kempf for their help with our project. We are also grateful to the two reviewers of this report, Michael Hansen from RAND and Paul Hogan from the Lewin Group, and to Aaron Kofner, who reviewed the code in the appendixes.

Abbreviations

ADMF	Active Duty Master Files
AFQT	Armed Forces Qualification Test
ASEC	Annual Social and Economic Supplement
AVF	All-Volunteer Force
BAH	basic allowance for housing
BAS	basic allowance for subsistence
BLS	Bureau of Labor Statistics
BPI	basic pay index
CPS	Current Population Survey
DECI	Defense Employment Cost Index
DMDC	Defense Manpower Data Center
DoD	U.S. Department of Defense
ECI	Employment Cost Index
FY	fiscal year
IPUMS	Integrated Public Use Microdata Series
NBER	National Bureau of Economic Research
NDAA	National Defense Authorization Act
ORG	Outgoing Rotation Group
OSD	Office of the Secretary of Defense
OUSD(P&R)	Office of the Under Secretary of Defense for Personnel and Readiness
QRMC	Quadrennial Review of Military Compensation
RMC	regular military compensation
SOFS	Status of Forces Surveys
SRB	Selective Reenlistment Bonus

| YATS | Youth Attitude Tracking Survey |
| YOS | years of service |

CHAPTER ONE

Introduction

In the all-volunteer military, pay is a key policy tool for recruiting and retaining personnel. Military pay must be high enough to attract and retain the personnel needed to meet requirements, and one measure of pay adequacy is how military pay compares to the pay of civilians with similar characteristics. Currently, military pay exceeds the civilian pay benchmark set by the Ninth Quadrennial Review of Military Compensation (9th QRMC) in the early 2000s for judging the adequacy of military pay and has exceeded that benchmark for a number of years, as discussed below. That benchmark is the 70th percentile of the pay of civilians with characteristics comparable to those of military personnel. Furthermore, since the 7th QRMC in 1992 and analysis by Hosek et al. (1992), it has been recognized that the Employment Cost Index (ECI), the index used by Congress to guide decisions about the annual adjustment to military basic pay, is constructed using data on the pay of civilians who have characteristics that are markedly different from those of military personnel. This report summarizes analysis performed for the 13th QRMC on setting the level and annual adjustment of military pay. Specifically, the study has two elements. The first provides an assessment of the continued relevance of the 70th percentile as a benchmark for setting the level of military pay. The second provides an assessment of the advisability of continuing to use the ECI to guide the annual military pay adjustment.

Background on RMC Percentiles and the ECI

Regular military compensation (RMC) is the measure of military pay that is considered most comparable to civilian pay.[1] RMC includes basic pay, basic allowance for housing (BAH), basic allowance for subsistence (BAS), and the federal tax advantage arising from allowances being tax-free. RMC accounts for approximately 90 percent of current cash compensation, and basic pay accounts for about 60 percent of RMC (Asch, Hosek, and Martin, 2002; Asch, 2019a). Comparisons of military and civilian pay generally start with a comparison of RMC with the full-time, full-year pay of comparable civilians. Deliberations of the President's Commission on an All-Volunteer Force in the 1970s noted that comparisons of military and civilian pay should recognize that military pay may need to be higher than that of comparable civilians because of the hazards and other conditions of military service (U.S. Department of Defense [DoD], 1970). Thus, the ability to meet recruiting and retention targets in light of the nature

---

[1] The Gorham Commission in 1962 developed the construct of RMC as a benchmark for comparing military compensation to civilian compensation, comprising of basic pay, BAH, and BAS. Later, the definition of RMC also included the federal tax advantage associated with receiving BAH and BAS tax-free.

of military service and the overall health of the All-Volunteer Force should be incorporated into analysis of the adequacy of military compensation, according to this Presidential Commission.

As mentioned, the 9th QRMC set the benchmark that RMC should reach at least the 70th percentile of the earnings of comparable civilians based on the input of past commissions and study groups, as well as studies of military pay and recruiting and retention outcomes (DoD, 2002). In particular, the 9th QRMC stated that,

> Military and civilian pay comparability is critical to the success of the All-Volunteer Force. Military pay must be set at a level that takes into account the special demands associated with military life and should be set above average pay in the private sector. Pay at around the 70th percentile of *comparably educated civilians* [italics added] has been necessary to enable the military to recruit and retain the quantity and quality of personnel it requires. (DoD, 2002)

From the standpoint of recruiting, DoD sets requirements for recruit quality. In particular, the benchmark for recruit quality is that 60 percent of recruits score in the top half of the distribution of aptitude scores on the Armed Forces Qualification Test (AFQT) and 90 percent are high school diploma graduates.[2] As discussed by Sellman (2004), DoD chose these benchmarks as the minimum acceptable values based on the 1990–1991 enlistment cohort and research that showed that high school graduates are less likely to leave before completing their enlistment contract and that higher-aptitude personnel perform better on military-related tasks.[3] This cohort was the group that produced satisfactory performance during Operations Desert Storm and Desert Shield. Analysis of military and civilian pay in the 1990s argued that pay at around the 70th percentile had historically been necessary to meet the DoD recruiting benchmark (Asch, Hosek, and Warner, 2001; Hosek and Sharp, 2001). For example, Hosek and Sharp (2001) showed that junior enlisted and officer pay during the 1990s were at about the 70th percentile of comparable civilian pay between 1993 and 1999. While the education benchmark is stated in terms of high school diploma status, the 9th QRMC concluded that the appropriate comparison group for enlisted personnel was no longer just those with a high school diploma but also those with some college. Tabulations by the 9th QRMC found that the military recruited from the college-bound youth market and that a large fraction of the enlisted force has some college attainment.[4]

---

[2]   AFQT scores are binned into groups or categories. Recruits with scores in categories 1 to 3A are in the top half of the AFQT distribution, where the scores are normed from 1 to 99 based on aptitude test scores for the American young adult population. High-aptitude recruits are therefore called recruits in AFQT Cat I–IIIA.

[3]   A large study, called "Project A," was conducted to validate the relationship between AFQT and hands-on military performance. Other research showed lower attrition among high school diploma graduates. See Sellman (1997, 2004), Buddin (1984, 1988), Green and Mavor (1994), Mayberry (1997), Orvis, Childress, and Polich (1992), "Project A" (1990), and Smith and Hogan (1994).

[4]   Another reason for setting military pay above civilian pay is that the DoD benchmarks require that the services select above-average youth in terms of high school graduate status and aptitude. Research has found that civilians with higher cognitive performance and better education earn more over their careers (Lin, Lutter, and Ruhm, 2018). Consequently, military pay would need to exceed the pay of civilians with similar education, experience, and gender to achieve desired recruit quality objectives, all else equal. Although this point was not explicitly mentioned by the 9th QRMC or subsequent analyses of the RMC percentile, it has been implicit in the argument that pay be set high enough to achieve recruiting and retention outcomes.

The 9th QRMC's report supported several actions mandated by the National Defense Authorization Act (NDAA) for Fiscal Year (FY) 2000. The 2000 NDAA authorized a 6.8 percent increase in basic pay in FY 2000 and basic pay increases equal to the percentage increase in the ECI plus one half of a percentage point through FY 2006. This legislation responded to a growing gap between military and civilian pay for mid-grade enlisted forces during the 1990s and to recruiting and retention problems in the late 1990s as a result of the "dot-com" boom in the civilian economy at that time. Congress also made several structural adjustments to the basic-pay table in the early 2000s, with targeted pay raises in grades E-5 through E-7 in July 2001, for example.

Following the terrorist attacks of September 11, 2001, and subsequent U.S. military operations in Iraq and Afghanistan, the basic pay increases of ECI + 0.5 percentage point were continued through fiscal year 2010 as insurance against a decline in either the size or quality of the military workforce resulting from the stress on the force due to frequent, long, and hazardous deployments. In addition to the higher-than-usual increases in basic pay, the housing allowance was raised in the early part of the decade to cover the full expected cost of off-base housing. Together, these pay actions succeeded in increasing basic pay and regular military compensation relative to civilian pay.

Reflecting in part the pay actions in the 2000s, the 11th QRMC found, in 2009, a decade after the 9th QRMC, that RMC was at about the 90th percentile of civilian wages for enlisted personnel.[5] To arrive at the 90th percentile for enlisted personnel, the 11th QRMC focused on civilians with high school diplomas, those with some college, and those with associate's degrees. For officers, the comparison groups were those with four-year college degrees and those with master's degrees or higher, and RMC was "at about the 83rd percentile" for these groups combined (DoD, 2012). The increase in RMC percentile between the 9th QRMC and the 11th also came from a downward trend in civilian wages between 2000 and 2009, which dropped 4 to 8 percent between 2000 and 2009 for most age and educational groups (Hosek, Asch, and Mattock, 2012).

Military pay growth slowed after 2010. The annual increase in military basic pay was set equal to the change in the ECI in 2011–2013 and was set below the ECI for 2014–2016. Since 2017, the annual basic pay increase has been set equal to the ECI. Regarding the change in civilian pay, the drop in civilian pay leveled out in 2012 and began rising in 2013, but in early 2008, civilian employment prospects fell when the unemployment rate rose precipitously. The unemployment rate increased for all education levels, and the number of workers employed part-time increased.

Recent RAND analysis has considered how military and civilian pay compared in more recent years (Smith, Asch, and Mattock, 2020). Our study of RMC percentiles for 2017 found that RMC was at the 85th percentile of the civilian wage distribution for enlisted personnel and at the 77th percentile of the civilian wage distribution for officers (Smith, Asch, and Mattock, 2020). These figures are somewhat less than the 90th percentile reported by the 11th QRMC for enlisted personnel and the 83rd percentile reported for officers, but we used a different methodology than the 11th QRMC. When using a comparable method, we found

---

5 The 9th QRMC compared RMC with the civilian wages of men, while the 11th QRMC used a weighted average of the wages of men and women, with the weights reflecting the gender mix in the military. In this report, we use a gender weighting. However, we also made pay comparisons separately by gender and found that RMC percentiles weighted by gender were only slightly higher than those for men only.

that RMC percentiles for 2009 were the same as for 2017, i.e., enlisted RMC was also at around the 84th percentile in 2009, similar to our estimate for 2017. Put differently, enlisted RMC relative to civilian pay remained unchanged between 2009 and 2017, and the differences from what was reported by the 11th QRMC are attributable to differences in methodologies.[6] Thus, despite the slowing of military pay growth after 2010, slowing in civilian pay was sufficient so that the comparison of military and civilian pay in 2017 was nearly identical to the comparison in 2009.

In short, analysis of the levels of military pay relative to the pay of comparable civilians indicate that the RMC percentiles now exceed the 70th percentile benchmark and have done so since at least 2009. As has been argued by the various QRMCs, as well as past studies, comparisons of military and civilian pay should also incorporate information on what it takes to ensure the successful recruitment and retention of the quality and quantity of personnel required by the armed services. Thus, assessment of pay adequacy needs to consider not only pay comparisons, including how military pay compares with the 70th percentile of comparable civilians, but also recruiting and retention outcomes. As shown in Smith, Asch, and Mattock (2020) and discussed in Hosek et al. (2018), measures of recruit quality, a key metric of successful recruitment, have greatly exceeded the DoD benchmarks for recruit quality for all services except the Army since 2009. For the Army, both the percentage of recruits in AFQT categories I–IIIA and the percentage who are high school diploma graduates have equaled or exceeded the DoD benchmarks since 2009. Another key metric is whether the services met their overall accession objective for a given year. Again, all of the services but the Army have met their annual accession objective in every year since 2009. In 2018, a year in which the RMC percentile greatly exceeded the 70th percentile, the Army fell short of reaching its accession mission. However, the Army has met its objective in every other year since 2009. A key question is whether an RMC percentile exceeding the 70th percentile is now needed for the services to meet their recruiting and retention objectives. The first part of our study focuses on this question.

Closely connected to the level of military pay is the methodology for adjusting basic pay each year. Section 1009c of Title 37 of the U.S. Code (2003) provides a formula for the annual increase in basic pay that is indexed to the annual increase in the ECI for the wage and salary of private-industry workers. The military pay increase is measured as the 12-month percentage change in the ECI for the third quarter of the calendar year in which previous years' values are used. For example, the ECI guiding the 2020 pay raise is the percentage change for the third quarter of 2018 relative to the third quarter of 2017. That said, the statute also allows the President to specify an alternative pay adjustment, so that the ECI is ultimately only a guide for adjusting basic pay. In particular, recruiting and retention outcomes will also inform the recommended pay raise, but they are also affected by the BAH and BAS components of RMC, not just basic pay. As discussed above, historically, the military pay increase has often deviated from the ECI increase.

The ECI is designed to track changes in the cost of labor over time while holding fixed the number of workers in each narrowly defined occupational category. The data for the ECI come from the National Compensation Survey, a survey of establishments. Thus, the ECI is

---

[6] The differences in methodology are discussed in Hosek et al. (2018) and are due to the method used to calculate years of experience, and weighting wages by the civilian distribution of educational attainment rather than the military distribution of educational attainment.

designed to be an index of employment cost for a given bundle of labor. It is not designed to track the median (or average) wage among civilian workers in the labor force.

Using the ECI to compare military with civilian pay growth has four advantages, as outlined by the 7th QRMC. These are its timeliness (the ECI is a quarterly indicator, allowing decisionmakers to use data collected three to four weeks prior to deliberating on the annual pay raise), its accuracy and stability in representing economy-wide compensation costs, the fact that it is produced by the Bureau of Labor Statistics (BLS) at no cost to the military, and its longtime use in guiding General Schedule scale pay increases. But to function well as an adjustment mechanism, the ECI should be relevant to military enlistment and reenlistment decisions. Hosek et al. (1992)—hereafter "Hosek 1992"—analyzed data between 1982 and 1991 and found that, although military pay lagged substantially behind the growth in the ECI, the services recorded no major recruiting and retention problems during this period. That is, measuring the "pay gap" using the ECI did not perform well in terms of tracking force-management outcomes.

The problem with the ECI is that it does not track well the opportunity wages that are relevant to military personnel. The ECI uses data on private-industry wage and salary workers who are, on average, older and who have a different mix of education and occupations than military personnel. Other pay-adjustment mechanisms that better reflect the demographics of military personnel could be used. Hosek 1992 developed the Defense Employment Cost Index (DECI) for this purpose, and analysis of pay-gap comparisons using the DECI found that the DECI tracked military enlistment and reenlistment more accurately than the ECI. This makes sense because the DECI is more focused on providing a representation of the opportunity wage of military personnel. In particular, two advantages of the DECI are that it uses individual earnings data from workers, rather than compensation data from business establishments, and it weights these earnings according to the demographic makeup of military personnel, which differs from the overall labor market in important respects. The latter feature makes the DECI a potentially more accurate reflection of the civilian earnings opportunities relevant to active duty service members.

That said, the ECI is available quarterly, whereas the DECI constructed by Hosek and Totten (2002) was an annual metric. The Hosek and Totten study also found that the ECI was less influenced by the more variable wages of young workers than was DECI, so it was more stable over time. However, this stability comes at the expense of a lack of sensitivity to the variability of opportunity wages relevant to military personnel over the business cycle.

When it was proposed in 1992, the DECI did not gain acceptance from DoD or Congress, and the 7th QRMC rejected the DECI as a replacement for the ECI, though it did recommend that the Office of the Under Secretary of Defense for Personnel and Readiness (OUSD[P&R]) underwrite further development of the DECI as a candidate index for future use in pay adjustment. Thus, the DECI was never adopted. But the ECI has clear disadvantages. The functioning of the ECI as an appropriate pay-adjustment mechanism, and alternative approaches, including the DECI, should be reassessed with more-recent data. The second objective of this study is to provide such an assessment in support of the 13th QRMC.

Research Questions and Approach

To meet the two objectives of the study, we address the following research questions:

1. How have recruiting and retention outcomes changed, and what changes have occurred in factors affecting recruiting and retention that would argue for a different benchmark than the 70th percentile?
2. How has military pay compared with civilian pay since the 1990s, when the 70th percentile benchmark was set?
3. How has military pay growth compared with civilian pay growth using the DECI, and does the DECI perform better than the ECI in terms of tracking recruiting and retention outcomes?
4. What are the policy implications of the analyses? Has the RMC percentile needed to be higher than the 70th percentile to meet recruiting and retention objectives? Does the 70th percentile benchmark need to be increased? What are the advantages and disadvantages of the DECI versus the ECI, and are the early criticisms of the DECI still relevant?

To address the first question, we provide an overview of how recruiting and retention outcomes have changed since the 1980s and 1990s, drawing on data provided by the OUSD(P&R) as well as military recruiting data and personnel data from the Defense Manpower Data Center (DMDC). We also use data from the Current Population Survey (CPS), a monthly survey of the U.S. population conducted by the BLS within the US. Department of Labor. We draw from past studies to identify other factors shown to affect recruiting and retention outcomes and draw on multiple data sources, including DoD data, to describe how these factors have changed over time. These factors include the unemployment rate and the state of the civilian economy, military propensity to enlist, end-strength requirements, and frequency of deployment.

To address the second question, we use military pay data from DoD's *Selected Military Compensation Tables* (OUSD[P&R], Directorate of Compensation, 1980–2018), also known as the Greenbook, and from Active Duty Pay Files provided by DMDC. For civilian pay, we use data from the CPS. To incorporate how the education and gender of military personnel have changed over time, we use data from the DoD Status of Forces Surveys (SOFSs) and from DMDC's Active Duty Master Files (ADMF). We weight civilian workers by the military gender mix then compute a civilian wage distribution for each level of education to make results comparable with previous QRMCs and studies.[7] Treating RMC as though it were a wage, we then find its placement in the distribution (i.e., we determine its percentile). We compute RMC percentiles for officers and enlisted personnel by years of service, as well as overall RMC percentiles for officers and enlisted personnel over time. Because the education weights in computing RMC percentiles differ depending on the data source—ADMF or SOFS data—we show RMC percentile results using each data source.

---

[7]   We do not weight by race, as our results would not be directly comparable with the results of studies in support of previous QRMCs. However, military pay does not differ by race, and the military tends to be more diverse than the civilian population. Thus, if we were to weight the civilian population by the military racial mix, our RMC percentiles would likely be higher than they presently are. In this way, our results are conservative estimates (that is, they are likely biased downward).

For the third question, we follow the Hosek 1992 methodology for computing the DECI using the March CPS and weights derived from the DMDC ADMF. This allows us to compare and validate our implementation of this approach relative to the earlier study and compute the DECI, ECI, and an index of basic pay from the early 1980s to the present using the original methods. We then build on and extend the earlier methodology to compute the DECI using the monthly CPS data for the Outgoing Rotation Group (ORG) rather than the March CPS and use a more refined version of the education weights from the ADMF. The use of the ORG subgroup allows us to compute the DECI with a shorter lag than the Hosek 1992 study. We compare the change in DECI with the change in the ECI and basic pay across time using the updated methodology. As with the Hosek 1992 study, we also consider DECI computations by subgroups, such as enlisted versus officers and by age and educational attainment and we consider how the DECI results and comparisons with the changes in the ECI and basic pay would change if we use SOFS data to derive weights, for more recent years when survey data are available. Finally, we examine how well the DECI tracks changes in recruiting and retention outcomes relative to the ECI.

We bring together these analyses to address the final questions related to the policy implications of the analyses.

Organization of This Report

In the next chapter, we present our overview of how recruiting and retention outcomes and the factors affecting those outcomes have changed. Chapter Three presents our analysis of the RMC percentile and shows how military pay has compared with civilian pay since the 1990s when the 70th percentile was set. In Chapter Four, we discuss our analysis of the DECI. We discuss policy implications in Chapter Five.

CHAPTER TWO

Changes over Time in Recruiting and Retention Outcomes and the Factors Affecting Them

In this chapter, we review broad trends in recruiting and retention outcomes and the factors that are correlated with those outcomes. The purpose of this review is threefold:

1. Provide contextual background: Trends in recruiting and retention outcomes are integral to any discussion about the appropriate level and annual adjustment to military pay. Further, trends in factors other than pay that influence outcomes are also relevant, especially since policy tools other than pay, such as bonuses and recruiters, can be used by the services to improve outcomes.
2. Assess changes relative to the 1990s: In considering trends, of particular interest is whether recruiting and retention outcomes and the factors correlated with those outcomes have changed relative to the 1990s when the 70th percentile was deemed adequate for setting military pay.
3. Assess changes relative to other possible benchmark years: While the 1990s as a benchmark period is an obvious choice, other periods might be relevant. Our objective is to consider alternative periods and then assess changes in recruiting and retention outcomes and the factors that are correlated with those outcomes in these other periods.

We begin the chapter with a discussion of periods other than the 1990s that might be considered when benchmarking military pay from the standpoint of DoD meeting its recruiting and retention requirements. We begin with this discussion, rather than end with it, so that we call out these specific periods when we show trends later in the chapter. Next, we review the trends in recruiting and retention outcomes and then review factors related to these outcomes. To provide historical context, we show and discuss the data from the early years of the All-Volunteer Force (AVF), to the extent the data are available (relevant to [1] above), and data from the benchmark years, relevant to (2) and (3) above. We conclude the chapter with a summary of findings.

Two aspects of the discussion are noteworthy. First, military and civilian pay are two factors correlated with recruiting and retention outcomes. But because we discuss pay in Chapters Three and Four, we do not discuss them in this chapter. Second, we note that four recent studies have shown trends over time in recruiting and retention outcomes and the factors related to these outcomes (Hosek, Asch, and Mattock, 2012; Hosek et al., 2018; Asch, 2019a; Smith, Asch, and Mattock, 2020). This chapter draws from these studies but also updates some of the information in them, both in terms of providing more recent data as well as extending trends farther back in time.

Considering the Benchmark Period

Chapter One provided a discussion of the origins of the 70th percentile and the relevance of the 1990s for using the 70th percentile benchmark for setting the level of military pay. But there may be another period or periods to consider as the benchmarks in terms of DoD meeting its recruiting and retention requirements, and we consider some alternatives in this section. Many changes relevant to recruiting and retention occurred during the 1990s. These included the Gulf War in 1990–1991, the drawdown of military end strength following the end of the Cold War in the early 1990s, operations in Bosnia and Kosovo in the mid-1990s, and the dot-com boom in the late 1990s, which saw a dramatic increase in the civilian pay of better-educated workers. For that reason, it is difficult to pinpoint exactly what part of the 1990s is the right year for a benchmark for comparing more recent outcomes and factors.

Ideally, we would use the RMC percentile for a period or group of periods when recruiting and retention were deemed adequate and efficiently achieved. Unfortunately, there has been no "steady state" for the military or for recruiting and retention over the past three decades, and hence no ideal period for us to select. Like the 1990s, the 2000s were a period of major change, including the attacks of 9/11, the operations in Iraq and Afghanistan and the large number of deployments to support those operations, and the Great Recession, which affected the civilian opportunities of potential recruits as well as military personnel. Recruiting and retention outcomes were deemed adequate for the most part during the 2010s. The exception was the Army, which missed its recruiting goal in 2018. These outcomes raise the question that motivates our study: Is the RMC percentile too high today? If so, then the satisfactory recruiting and retention outcomes that were achieved during the 2010s may have been achieved inefficiently. For that reason, the 2010s are not an ideal period either.

Because no single period is ideal, our analysis considers four comparison or benchmark periods:

1. 1993–1997
2. 1988–1989
3. 2010
4. 2011–2013.

The period 1993–1997 roughly corresponds to the data used by past studies, discussed in Chapter One, to show that military pay was around the 70th percentile of pay for similar civilians and the period around which the Office of the Secretary of Defense (OSD) considered recruiting and retention outcomes to be adequate. But because of the large changes during the 1990s and because DoD considered the force that was deployed to the 1990–1991 Gulf War as sufficiently high-quality, we also consider the period 1988–1989. The force that was deployed to the Gulf War was recruited and retained in the late 1980s, and, by considering the late 1980s, we avoid including the early 1990s, when a short recession occurred.

The third period we use as a benchmark is 2010, a year that followed the 2006–2008 surge of forces in overseas operations, though overseas deployments were still extensive. DoD end strength also began to decline after 2010 along with accession goals, so 2010 marked a transition year from the wartime footing of the previous eight years. The year 2010 was also the height of the Great Recession, and, as might be expected given research showing that recruiting and retention outcomes improve with increases in the unemployment rate, 2010 was a year when recruiting and retention outcomes were deemed satisfactory by DoD (Hosek, Asch, and Mattock, 2012).

The final period we use is 2011–2013. These years represent the period prior to the slow-ing of the growth of military pay. The slowing of pay growth after 2013 occurred because it was deemed that recruiting and retention were in good shape and the United States had ended the war in Iraq and planned to reduce its presence in Afghanistan.

As we discuss trends in recruiting and retention outcomes in the next section and trends in factors related to these outcomes in the section after that, we first provide a general overview of the trends for contextual background and then discuss how recent outcomes in 2018 and 2019 compare with those in the benchmark years.

Recruiting and Retention Outcomes

Recruiting Goals and Quality: Contextual Background

Figure 2.1 shows recruiting goals for each service and for DoD since 1980. The most striking feature of the trend over time is that the overall accession goal of 171,067 in 2019 was less than half of what it was in 1980, owing in large part to the defense drawdown in the early 1990s, during which accession goals for DoD dropped from 292,021 in 1989 to 174,806 by 1995. Since 2000, recruiting goals for DoD decreased on net, but the timing of the decrease dif-fered by service, and in fact some of the services, notably the Army and Marine Corps, saw an increase in accession goals associated with operations in Iraq and Afghanistan. After dropping from 2008 to 2013, recruiting goals for DoD overall have increased.

Figures 2.2 and 2.3 show the trends in the quality of non–prior service recruits since the beginning of the AVF in 1973: Figure 2.2 shows the percentage of recruits with at least a high school diploma (Tier 1 recruits[1]), and Figure 2.3 shows the percentage in AFQT categories I–IIIA. It can be seen from Figure 2.2 that quality was particularly low in the early years of the AVF in the late 1970s. As shown by Hosek, Petersen, and Heilbrunn (1994), military pay lagged behind civilian pay in the late 1970s. Furthermore, AFQT scores were misnormed,[2] and DoD inadvertently enlisted a large number of lower-aptitude recruits. Furthermore, the Vietnam-era GI Bill was replaced with the substantially less generous Veteran's Assistance Education Program in 1976, resulting in a large decline in enlistment supply (Goldberg, 1982). Lagging military pay along with poor recruit quality led Congress to provide a 25 percent increase in military pay over the years 1981 and 1982. Since the mid-1980s, the percentage of

---

[1]   A Tier 1 recruit is one who is a high school graduate, has an adult-education diploma, or has completed at least one semester of college or attended virtual or distance learning or an adult or alternative school. See DoD, 2016, Appendix A, p. 13.

[2]   The AFQT is a composite of math and verbal scores from the Armed Forces Vocational Aptitude Battery, or ASVAB. When the ASVAB was implemented in the mid-1970s, there were undetected flaws in the methodology used to determine the AFQT percentiles in reference to the normative population, namely the United States young adult population. Be-cause of the misnorming, many recruits entered the military who would have otherwise been identified as below average in terms of AFQT scores. The flaw was detected and fixed in 1980. See Ramsberger et al. (1999).

Figure 2.1
Fiscal Year Enlisted Active Duty Accession Goals, by Service

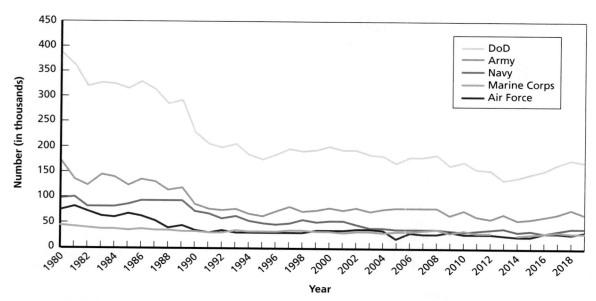

SOURCE: OUSD(P&R).
NOTE: Marine Corps data for 2013 are missing.

Figure 2.2
Percentage of Enlisted Accessions That Are Tier 1, by Service

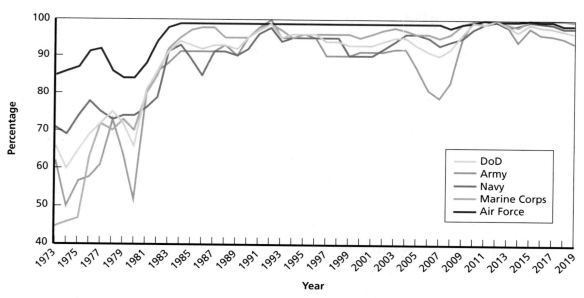

SOURCE: OUSD(P&R).

Tier 1 recruits has stayed at or above the DoD benchmark requiring that at least 90 percent of accessions have at least a high school diploma. The notable exception is the Army, for which the percentage fell below the Tier 1 benchmark between 2005 and 2009; it rebounded in 2010 and the Army has since exceeded the Tier 1 benchmark, though the percentage has declined for the Army since 2013, as it has for all of the services but the Air Force.

Figure 2.3 shows that the percentage of recruits in AFQT categories I–IIIA rebounded in the 1980s, as well, and has exceeded the DoD 60 percent benchmark for each of the services since then. The percentage rose in the late 1980s and early 1990s, reaching 72 percent in 1992, but then declined in the 1990s to 66 percent across DoD by 2000.

As military pay also increased faster than the ECI and RMC increased during the 2000s, recruit aptitude rose for all services but the Army. The Navy saw a relatively steady rise in percentage of recruits in AFQT categories I–IIIA between 2000 and 2010, reaching 83 percent of accessions by 2010, and the Air Force saw a particular bump up in aptitude between 2002 and 2006, reaching 91 percent of accessions by 2010. The rise over this period for the Marine Corps was interrupted with a bit of a drop between 2006 and 2008, during the surge in forces in Iraq and Afghanistan. Still, even during this period, the percentage of Marine Corps recruits in AFQT categories I–IIIA was 65 percent in 2007, above the DoD benchmark of 60 percent, and reached 73 percent by 2010. For the Army, recruit aptitude rose between 2000 and 2003, reaching 73 percent of accessions being in AFQT categories I–IIIA by 2003. But Army recruit aptitude declined thereafter, to 61 percent by 2006. Since, then, the percentage of Army accession in AFQT categories I–IIIA has hovered at just above 60 percent, the DoD benchmark. Thus, the Army pattern during the late 2000s in particular was markedly different than the other services. During the 2010s, recruit aptitude declined for all the services but the Army, but for the other services aptitude stayed well above the DoD benchmark. Again, Army recruit aptitude has continued to hover at the 60 percent DoD benchmark.

Figure 2.3
Percentage of Enlisted Accessions That Are in AFQT Categories I–IIIA, by Service

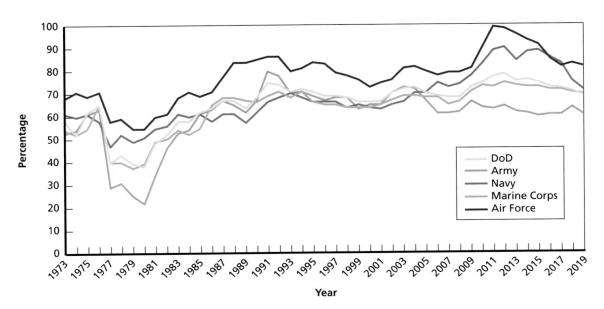

SOURCE: OUSD(P&R).

The reasons why the Army failed to increase its quality of recruits at a time of increasing basic pay and RMC, especially after 2003, is explored in Hosek et al. (2018, pp. 71–73). One reason may be that Army recruiting became more difficult than other services' recruiting. However, DoD Youth Polls from 2001 to 2015 show that the trend in youth interest in the military was quite similar across the services.[3] Another possibility is that the Army set quality goals and programmed recruiting resources to sustain, but not increase, accession quality, unlike the other services. That is, the Army set its recruiting-quality goals constant as military pay increased, allowing the Army to hold down recruiting resource costs such as for bonuses and advertising, yet still meet the DoD benchmark. The other services focused on increasing recruit quality above the benchmarks when military compensation increased.

Another possibility is that higher military pay might have affected recruiter effort. Research suggests that recruiters exert less effort when pay and other recruiting resources are plentiful (Dertouzos, 1985). As we show below, there is some evidence that recruiter productivity decreased after 2008, particularly for the Army, suggesting that Army recruiters might have reduced effort in the recession era. Yet another possibility is that the introduction of the post-9/11 G.I. Bill in 2009 eliminated the Army's ability to provide education benefit "kickers" to recruits entering selected occupations; these kickers gave the Army a recruiting advantage over the other services. Without this advantage, it may have been more difficult for the Army to expand high-quality recruiting relative to the other services because recruits in all services had access to the post-9/11 G.I. Bill.

Recruiting Goals and Quality: Comparisons Across Benchmark Years

We summarize in Table 2.1 how recruiting outcomes compare in recent years (the average of 2018–2019) with those in the benchmark years. (We discuss other elements in the table later in this chapter.) Because the pattern of the Army's recruit quality in recent years differed from that of the other services, we show statistics for all of DoD in the top panel of the table and for all of DoD except the Army in the lower part of the table.

With respect to recruiting and recruiting quality in particular, the percentage of Tier 1 recruits exceeded the 90 percent DoD benchmark in 2018–2019, and this was also the case in each of the earlier periods considered. For aptitude, the percentage of recruits in AFQT categories I–IIIA was 69 percent for all of DoD in 2018–2019, lower than percentages in the 2010 and 2011–2013 periods, when recruit aptitude was particularly high, reaching 77 percent in the latter period. But, the 69 percent figure for 2018–2019 was about the same as it was in 1993–1997, the period when some of the analyses that led to the 70th percentile were computed. This could suggest that recruit quality today is about the same as it was during the mid-1990s, and just above what it was in the late 1980s, when recruit aptitude was 66 percent in AFQT categories I–IIIA.

But the picture changes when we exclude the Army, as shown in the lower panel in Table 2.1. The percentage of recruits in AFQT categories I–IIIA in 2018–2019 across DoD (except the Army) was 75 percent, not 69 percent, and this figure exceeded the percentages in the mid-1990s (71 percent) and in the late 1980s (67 percent). As with the all-DoD cases, recruit aptitude in 2018–2019 was less than in 2010 (83 percent) or in 2011–2013 (86 percent), when we consider all of DoD but exclude the Army. Thus, recruit aptitude was higher in

---

[3]   The DoD Youth Polls are discussed later in this chapter.

Table 2.1
DoD Recruiting and Retention Outcomes and Selected Factors Related to Outcomes, 2018–2019 and Selected Benchmark Years

	Average 2018–2019	Average 2011–2013	2010	Average 1993–1997	Average 1988–1989
All Services					
Recruiting					
Accession mission	171,155	160,099	165,362	189,975	290,343
Percentage Tier 1	96.9	99.1	99.2	95.4	92.5
Percentage AFQT Categories I–IIIA	69.0	77.0	74.4	70.4	65.5
Retention					
Enlisted continuation rate at YOS 4	73%	69%	72%	63%	62%
Enlisted continuation rate at YOS 8	85%	82%	86%	84%	87%
Officer continuation rate at YOS 8	91%	92%	94%	89%	88%
Adult unemployment rate	3.8	8.1	9.6	5.78	5.4
Military propensity	13.0	13.1	12.5	14.2	18
Enlisted end strength	1,083,131	1,134,275	1,164,553	1,277,637	1,815,034
Deployments	17,370	168,525	257,674	41,770	5,947
Recruiters	14,367	13,589	14,627	11,967	12,796
Accessions per recruiter	11.2	11.8	11.3	15.9	22.7
Enlistment bonuses ($1,000)[a]	$530,191	$343,385	$701,581	$48,357	$113,369
Reenlistment bonuses ($1,000)[a]	$1,092,816	$843,158	$1,030,551	$386,126	$865,952
All Services Except Army					
Recruiting					
Accession mission	102,076	95,545	90,762	117,167	172,383
Percentage Tier 1	98.6	99.3	98.8	96.2	93.5
Percent AFQT Categories I–IIIA	75.0	86.2	82.5	70.9	66.9
Retention					
Enlisted continuation rate at YOS 4	76%	71%	72%	58%	64%
Enlisted continuation rate at YOS 8	85%	83%	85%	86%	87%
Officer continuation rate at YOS 8	93%	92%	93%	87%	88%
Enlisted end strength	699,828	703,085	714,233	844,998	1,156,030
Deployments	8,154	168,525	95,858	29,490	1,261
Recruiters	6,609	7,218	8,035	7,047	7,229
Accessions per recruiter	15.5	13.3	11.3	16.7	23.9
Enlistment bonuses ($1,000)[a]	$109,584	$88,327	$174,522	$19,903	$40,197
Reenlistment bonuses ($1,000)[a]	$639,936	$626,651	$778,097	$312,807	$673,932

SOURCE: OUSD(P&R), and tabulations provided by DMDC.

NOTE: YOS = years of service.

[a] Constant 2019 dollars. Bonus totals exclude Air Force.

2018–2019 than in the early benchmark years of the late 1980s and mid-1990s but lower than in the 2010 or 2011–2013 periods.

Continuation

As indicators of retention, we consider one-year continuation rates for enlisted personnel at 4 and at 8 years of service (YOS) and for officers at 8 YOS. Continuation rates are not the same as retention rates because the latter is the percentage of personnel who are eligible to leave (for example, because they have completed a service obligation) who choose to stay in service. Continuation rates are the percentage of personnel on hand at the beginning of the year who are still on hand at the end of the year. Some of these personnel are still under a service obligation. Furthermore, if the services change service obligation contract lengths (making them shorter or longer), continuation rates at a given point in a career, such as YOS 4, could appear higher or lower, simply by virtue of the service policy and not because service members have a higher or lower propensity to stay in service, all else equal. More generally, the observed continuation rates reflect service personnel policies rather than voluntary retention behavior. For example, when a service is reducing end strength, it might tighten eligibility for retention, thereby reducing continuation rates. For that reason, the continuation rates we show should be considered rough indicators of voluntary retention behavior. We approximate enlisted first-term and second-term retention with continuation rates at YOS 4 and YOS 8, respectively, and officer retention in the early mid-career at YOS 8.

Table 2.1 shows continuation rates for all of DoD and for DoD except the Army. Two broad trends are apparent. Relative to the late 1980s and mid-1990s, retention in 2018–2019 was higher for first-term enlisted personnel and for officers. For example, average YOS 4 enlisted continuation rates were 62 percent in 1988–1989, but 73 percent in 2018–2019. The differences between the early periods and the most recent period are a bit larger when the Army is excluded: 76 percent in 2018–2019 versus 64 percent in 1988–1989 for first-term enlisted personnel. Second-term retention, as proxied by the YOS 8 continuation rates for enlisted personnel, has been relatively more stable over time across DoD, especially when the computation excludes the Army. Overall, across DoD, retention has broadly been sustained or increased in 2018–2019 relative to the earlier benchmark periods.

That said, the continuation rates in Table 2.1 mask considerable variation over time within services and across services. For example, Army first-term retention, as measured by YOS 4 continuation, was roughly the same between 1995 and 2006, with an increase and then a subsequent decrease between 2006 and 2013, and has been generally stable since 2013. In contrast, Air Force first-term retention fell sharply between 1995 and 2000 but has increased since then, with a particularly sharp increase between 2000 and 2002. Trends over time by service are shown in Appendix A.

Factors Related to Recruiting and Retention Outcomes

A large body of research has analyzed empirically the factors associated with recruiting and retention outcomes,[4] and three recent studies have provided overviews of recent trends in some of these factors (Hosek et al., 2018; Asch, 2019a; Smith, Asch, and Mattock, 2020). We update

---

[4] See Asch and Warner (2018) and Asch (2019b) for recent reviews of these studies.

trends presented in recent studies (shown in Appendix A) and show trends in other factors. We also show these factors in the most recent year, 2018–2019, and in the four previous benchmark periods.

Unemployment

When the economy worsens, workers have fewer outside options and may be more likely to be enticed to join the military, which provides relatively high and stable pay. Table 2.1 shows that the adult unemployment rate was substantially lower in 2018–2019 (3.8 percent) than in the previous periods (for example, 5.8 percent in the mid-1990s). The table also shows evidence of the Great Recession; in 2010, the adult unemployment rate was 9.6 percent.

End Strength

Another factor related to recruiting and retention outcomes is end strength, or how many people the services need. When the demand for forces increases or decreases, especially when changes occur rapidly, recruiting and retention may become more difficult or easier. Figure 2.4 show trends in enlisted and officer end strength. As with the trend in accessions, the figures show the drawdown of forces in the early 1990s at the end of the Cold War. Although there were periods of stability, the Air Force and Navy continued to reduce force size into the 2000s with an uptick after 2014. The Army and Marine Corps increased strength during the 2000s, especially Army officer strength, but decreased strength in the early 2010s. Since 2016, both Army and Marine Corps strength have increased. Despite these recent increases, DoD enlisted strength (with and without the Army included) was lower in 2018–2019 than in any of the earlier benchmark periods, as shown in Table 2.1.

Deployments

Another factor related to retention is the extent of long and hazardous deployments (Hosek and Totten, 2002; Fricker, Hosek, and Totten, 2003). As with previous studies, we use the number of personnel receiving imminent-danger or hostile-fire pay to measure deployment, with 2000 normalized to one for each service. Deployments during Operation Desert Storm in 1990–2002 were extensive, as were deployments during operations in Iraq and Afghanistan. But by 2015, with the drawdown of forces in Iraq and Afghanistan, deployments were a fraction of their levels in the late 2000s. This is seen in Table 2.1. The number of service members receiving imminent-danger or hostile-fire pay averaged 17,370 in 2018–2019, less than the number in any of the benchmark period, with the exception of 1988–1989, when the number receiving either pay was 5,947.

Propensity

To gauge youth interest in military service, DoD surveys American youth ages 16–21 and asks about their propensity to join the military. Propensity is measured by a four-category response to a question about how likely is it that the individual will join the military in the near future, where the possible responses are "definitely," "probably," "probably not," and "definitely not." Those who respond "definitely" or "probably" are deemed positively "propensed." Research shows that positively propensed youth have a higher likelihood of enlistment (Ford et al., 2009). Figure 2.5 shows the percentage of youth who stated a positive propensity, by year, since 1984. The blue and red lines show percentages from the Youth Attitude Tracking Survey

Figure 2.4
Enlisted and Officer End Strength, by Service

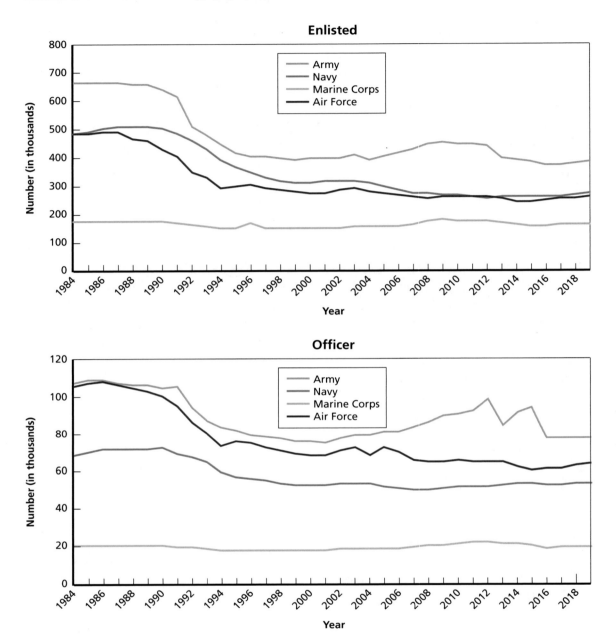

SOURCE: OUSD(P&R).

(YATS) and the Youth Poll, respectively. Because of differences in methodology, including frequency of administration, results from the two surveys are not directly comparable.

Propensity dropped dramatically from 18 percent in the late 1980s to about 13 to 15 percent in the 1990s. The Youth Poll began in April 2001 and showed that, following 9/11, youth propensity rose from 12 percent in April 2001 to 15 percent by November 2001, just nine months later. Propensity remained high through 2005 but dropped to 10 percent in 2006 at the beginning of the "surge" of forces but increased back to 12 percent by 2008. Since 2008,

Figure 2.5
Percentage of Young Adults Expressing a Positive Propensity to Serve in the Military

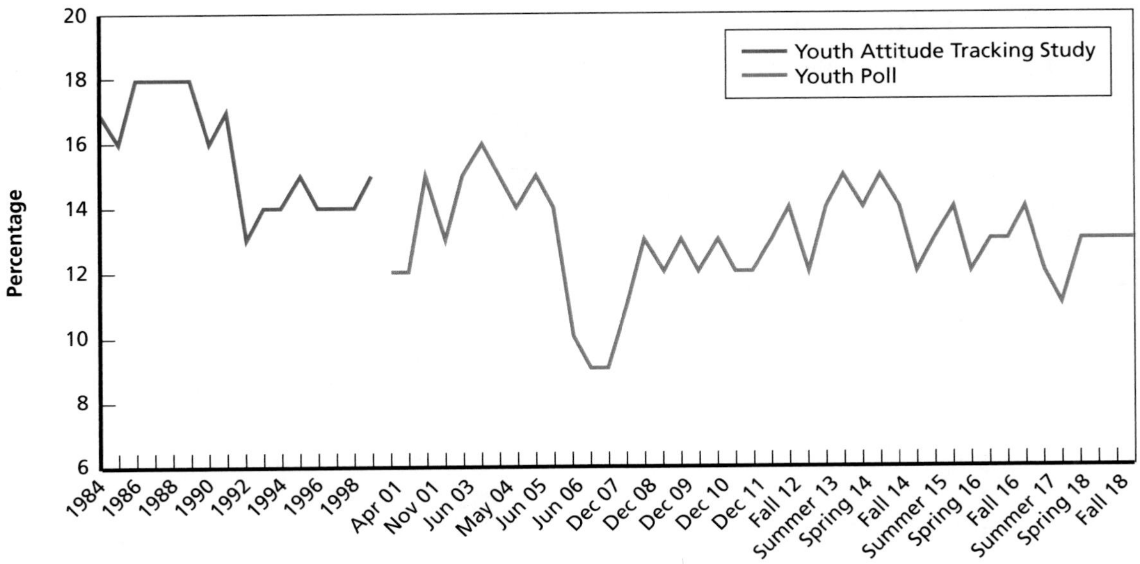

SOURCE: DoD, Office of People Analytics, Youth Attitude Tracking Study and Youth Polls, 1984–2018.
NOTE: Chart shows the percentage of 16–21-year-olds responding "definitely" or "probably" to the question "How likely is it that you will be serving in the Military in the next few years?" for December 2010–fall 2018 and to the question "Now, I'd like to ask you how likely is it that you will be serving in the Military in the next few years?" for 1984–June 2009.

the percentage of youth expressing a positive propensity has been stable at generally between 12 and 14 percent, though it reached 15 percent in fall 2013 and summer 2014.

Propensity rates since 2008 have been in the same range as or even somewhat higher than the rate in April 2001, just prior to 9/11, and at about the same range as propensity (as measured by YATS) during the 1990s, albeit sometimes lower. Given the dramatic changes in the military, its operations, and the civilian environment that it recruits from, the general stability of the rate of positive propensity, notwithstanding the variations already noted, is striking, especially since 2008, a period that covered the Great Recession and the drawdown of operations in Iraq and Afghanistan and the strong civilian economy since 2014. This stability, especially during tight labor markets, likely reflects the efforts on the part of services, OSD, and Congress to sustain recruiting. That is, the military pay raises that exceeded the ECI during the 2000s, along with increases in enlistment bonuses (see next subsection) and other efforts, helped sustain youth interest in the military. As evidence, Warner et al. (2001) and Warner et al. (2002) found that propensity to serve is positively related to military pay relative to civilian pay. Put differently, although propensity is used by DoD and policymakers to gauge underlying youth attitudes toward the military, it should be more accurately considered an indicator of youth interest, recognizing that youth interest reflects in part the efforts by DoD to influence that interest.

As shown in Table 2.1, the percentage of youth expressing a positive propensity to serve was 13 percent in 2018–2019, taken from the DoD Youth Polls. This figure is about the same as the percentages in the 2010 and 2011–2013 periods and slightly lower than the 14.2 percent average for 1993–1997 from YATS. It was substantially lower than the 18 percent average for the 1988–1989 period.

Recruiters and Recruiter Effort

Research shows that high-quality enlistments increase as the number of recruiters increase. In particular, studies typically estimate a recruiter elasticity of about 0.5, implying that high-quality enlistments increase by about 5 percent when the stock of Army recruiters increases by 10 percent.[5] Figure 2.6 shows the average monthly number of recruiters in each year, relative to 1989, for all of DoD and for all of DoD except the Army. We show both curves to highlight the difference in trends for the Army versus the rest of DoD in more recent years. The number of recruiters declined during the defense drawdown years of the early 1990s but increased in the mid-1990s as recruiting became more challenging during the dot-com boom. Recruiters continued to increase through 2002 and the invasion of Iraq but decreased through 2004. The surge of forces that began in 2006 saw an increase in the size of the recruiter force, peaking in 2009 at the start of the Great Recession.

Beginning in 2009, the number of recruiters declined until 2012 for all of DoD and until 2014 when the Army is excluded from the total. Since then, the number of recruiters has sharply increased across DoD, though the difference in trend when the Army is excluded suggests that the sharp increase after 2012 is primarily due to increases for the Army. We show service-specific trends in Figure A.6 in Appendix A. The figure shows the sharp increase in the size of the Army recruiter force since 2012, relative to the other services. It also shows that trends differed markedly across the services in general. For example, between 2000 and 2005,

Figure 2.6
Index of Annual Average Monthly Recruiter Count, DoD and DoD without the Army

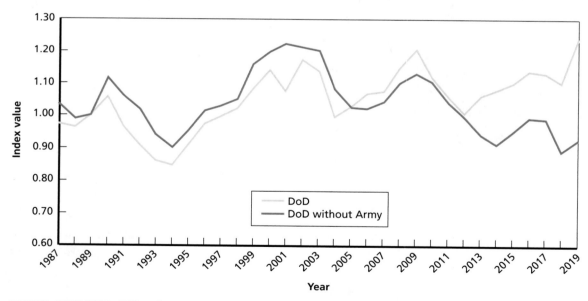

SOURCE: OUSD(P&R), Office of Accession Policy.
NOTE: The year 1989 = 1.00P.

---

[5]   A review of these studies is provided in Asch (2019a), Table 3.

the average number of recruiters in each year was relatively stable for the Army and Marine Corps, fell for the Navy, and increased for the Air Force.

The positive relationship between recruiters and enlistments is not automatic. Recruiters make up a workforce, and they are a human resource that must be properly managed to be effective and efficient. Research shows that one of the most important factors affecting recruiter productivity at the individual and recruiter station level is the mission assigned to that recruiter or station. Beyond mission, the other factors that explain variations in the number of high-quality enlistment contracts signed by a recruiting station are (1) the quality of the market in which the recruiter operates, (2) nationwide differences in the recruiting environment over time, (3) measured personal attributes of the recruiter, (4) station size, and (5) region of country (Dertouzos and Garber, 2006). Local market quality is the set of factors that capture the difficulty of making mission. These factors include local economic conditions, market demographics, and the size and age distribution of the veteran population.

To provide a general overview of how recruiter productivity has changed over time, Figure 2.7 shows annual accessions per recruiter for all of DoD and for all of DoD except the Army. Service-specific trends in accessions per recruiter are shown in Figure A.7 in Appendix A. Because recruiters are measured as an average monthly amount but accessions are measured annually, a value of 12 in the figures means that, on average, recruiters produced 12 accessions per year, or one accession per month. The most marked trend in the figure is the decline in accessions per recruiter since the late 1980s. In 1988, the average was 24, implying that, on average, recruiters produced two accessions per month. The yearly figure fell to about 15 by 1996 and to 12, or an average of 1 accession per month, by 2006. Recruiter productivity increased above 12 per year between 2006 and 2009, but then fell below 12 in 2010. Since 2010, accession per recruiter has remained at or below 12, though the trend for all of DoD shows some increase between 2010 and 2014 and again between 2015 and 2019. That said,

Figure 2.7
Annual Accessions per Recruiter, DoD and DoD without the Army

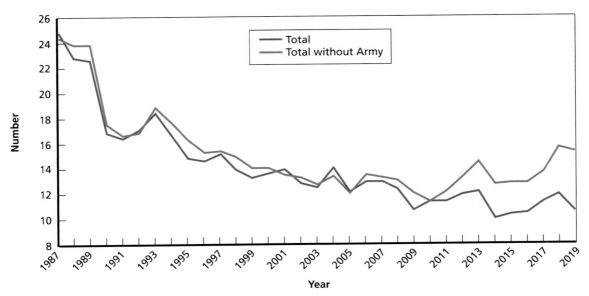

SOURCE: OUSD(P&R).

if the Army is excluded from the computation, Figure 2.7 shows that accessions per recruiter has risen since 2011, reaching about 15 by 2018, a figure that is comparable to the productivity figure in the mid-1990s. Figure A.7 shows that since 2011, accessions per recruiter have declined for the Army but increased for the other services.

Comparing accessions per recruiter in 2018 and 2019 versus the benchmark periods in Table 2.1 shows the decline in recruiter productivity relative to the 1980s but also shows that 2018–2019 productivity was similar to productivity in the mid-1990s, when the Army is excluded, and lower than the mid-1990s when the Army is included in the computations.

Enlistment Bonuses

The services also make use of enlistment bonuses as an incentive to expand enlistment supply and channel recruits into hard-to-fill skill areas or for longer enlistment terms. Estimates of the elasticity of high-quality enlistment with respect to expected bonus amount vary from 0.04 to 0.17.[6] These estimates imply that a doubling of the average enlistment bonus (a 100-percent increase) would expand high-quality enlistments by between 4 and 17 percent. Thus, the market-expansion effect of bonuses is relatively modest. An advantage of bonuses is that they can be deployed immediately. Funds can be reprogrammed within a fiscal year, and bonuses can be directed to recruits almost immediately. In contrast, recruiters need to be assigned and trained, and it takes time for them to reach full productivity. An advantage of enlistment bonuses over military pay is that they can be targeted to critical skill areas or longer term lengths.

Figure 2.8 shows the trends in enlistment bonus budgets (in constant 2019 dollars) relative to 1989 for the Army and Navy in the top panel and for the Marine Corps in the bottom panel.[7] Bonus budgets declined in the 1990s relative to 1989, so that by 1995, the Army and Navy budgets were about a quarter of what they were in 1989, and the Marine Corps budget was about 60 percent of its 1989 budget. Since the mid-1990s, enlistment bonus budgets have increased, rising especially sharply beginning in 2005 for the Army and Navy and in 2007 for the Marine Corps. For example, the Marine Corps enlistment bonus budget was 23 times as large in 2009 as it was in 1989, in constant dollars, and the Army budget was almost 10 times as large. Budgets were subsequently sharply reduced; by 2013, the Marine Corps budget was at the level it was in 2004 (or 3.28 times its 1989 level), the Navy budget was near its 1989 level, and the Army budget was near its level in 2000 (or 1.78 times its level in 1989). Since then, the Marine Corps budget has declined, and both the Army and Navy budgets have increased.

Reenlistment Bonuses

The services use selective reenlistment bonus (SRB) programs as an incentive to induce enlisted members to "re-up," or sign another enlistment contract, and SRBs are typically targeted to members in critical occupational specialties. The ability to target SRBs is an advantage over military pay. Research has consistently found that increases in SRBs increase retention, independent of the effects of military pay and the civilian economy (Hosek and Totten, 2002; Hogan et al., 2005; Asch et al., 2010; Asch and Warner, 2018). These studies typically use a "reduced-form" approach that focuses on estimating the effects of bonuses on the first-term

---

[6]   See Table 2 in Asch (2019a) for a summary of these studies.

[7]   We show the Marine Corps in a separate chart to allow for a different scale that accommodates the spike observed in 2009. Air Force data are not shown because of unreliable information.

Figure 2.8
Index of Enlistment Bonus Budget in Constant 2019 Dollars, by Service

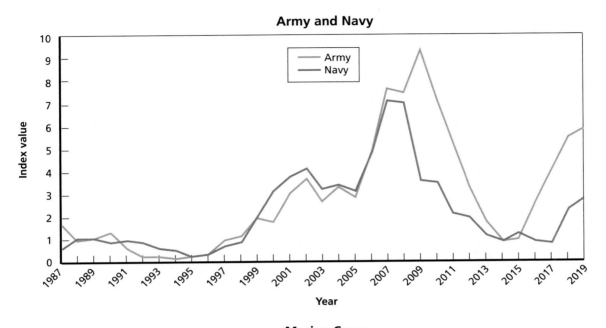

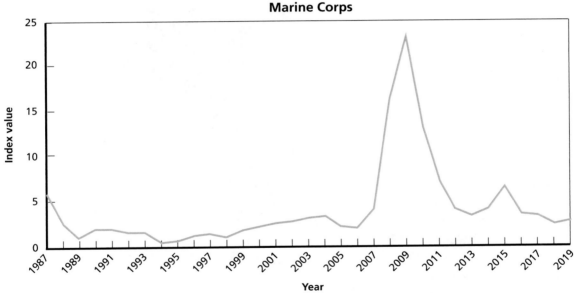

SOURCE: OUSD(P&R).
NOTE: The year 1989 = 1.00.

or second-term reenlistment rate. The formula for the size of an SRB is the product of term length (the years of additional service obligated by the new contract), monthly basic pay at the time of reenlistment, and the bonus multiplier.[8] Reduced-form models provide elasticity estimates of the effects on reenlistment of an increase in the SRB multiplier equal to 1 (or a one-

---

8   In 2009 the Army moved away from using multipliers and began using a system that gives lump sum payments based on grade, obligation length, and tier level, where the higher-tiered bonuses have higher dollar values.

step increase) and also include controls for other observable factors, including military pay and the civilian economy. As an example of results from a reduced-form model, Asch et al. (2010) found that a one-step increase in the SRB multiplier (which represents one month of basic pay per year of reenlistment) was estimated to increase the Army first-term reenlistment rate by about 3–4 percentage points. Navy first-term reenlistment was also estimated to rise by 2.5 percentage points per unit increase in the SRB multiplier; Marine Corps reenlistments were predicted to rise by 3.5 percentage points.

Figure 2.9 shows the trends in reenlistment bonus budgets (in constant 2019 dollars) relative to 1989 by service in the top panel and across the DoD in the bottom panel. Similar to enlistment bonus trends shown in Figure 2.8, SRB budgets declined in the 1990s relative to 1989 so that by 1995, the DoD budget was about 40 percent of its 1989 budget. The Air Force SRB budget increased markedly beginning in the late 1990s, and budgets for the Army and Marine Corps increased sharply in 2004. However, by 2011, Army and Marine Corps SRB budgets were comparable in real terms to their levels in the late 1980s. The Army budget again increased beginning in 2016. Across DoD (bottom panel), excluding the Army shows that DoD SRB budgets in recent years are comparable to their levels in the late 1980s.

Summary

In this chapter, we showed trends in recruiting and retention outcomes and the factors correlated with these outcomes, except military and civilian pay, two factors discussed in Chapter Three. Comparing the most recent outcomes for which we have data, 2018 and 2019, with those for 1988–1989 and 1993–1997—the periods that were relevant to the studies that led the 70th percentile being set as the benchmark for military pay levels—we find that, relative to the earlier periods, recruit quality across DoD, measured in terms of the percentage of accessions that are AFQT categories I–IIIA and are Tier 1, is about the same in recent years. However, when the Army is excluded from the measurement, recruit quality is higher in recent years. Furthermore, retention, measured in terms of the continuation rate at YOS 4 for enlisted personnel and at YOS 8, is also higher, regardless of whether the Army is excluded. As the other services increased recruit quality, the Army kept quality close to the DoD benchmarks for recruit quality, for reasons that are unknown. The higher quality of recruits and the higher continuation rates are notable because the adult unemployment rate was lower in 2018–2019 than in the earlier periods, implying that recruiting and retention likely were more difficult for the services to sustain. On the other hand, deployments in 2018–2019, while higher than deployments in the late 1980s and mid-1990s, were substantially lower than in the 2000s and early 2010s. The size of the recruiter force in 2018–2019 was somewhat higher than in the late 1980s and mid-1990s, though accessions per recruiter were lower, indicating that recruiters were less productive. One possible explanation for the lower productivity was less interest in the military on the part of American youth, especially in light of the historically low unemployment rate in the 2018–2019 period. However, the trend in youth propensity to enlist does not entirely support this explanation. Although proportion of youth expressing a positive propensity to enlist has varied over time, and recent estimates of propensity are not directly comparable to those from before 2001 because of changes in the survey methodology, available estimates indicate that propensity to enlist in 2018–2019 was only slightly lower than propensity in the mid-1990s, 13.0 percent in 2018 versus 14.2 percent on average between 1993–1997. On the other hand, the average propensity in the late 1980s was 18 percent, significantly higher than the 13.0 percent for the recent period.

Figure 2.9
Index of Reenlistment Bonus Budget in Constant 2019 Dollars, by Service and for DoD

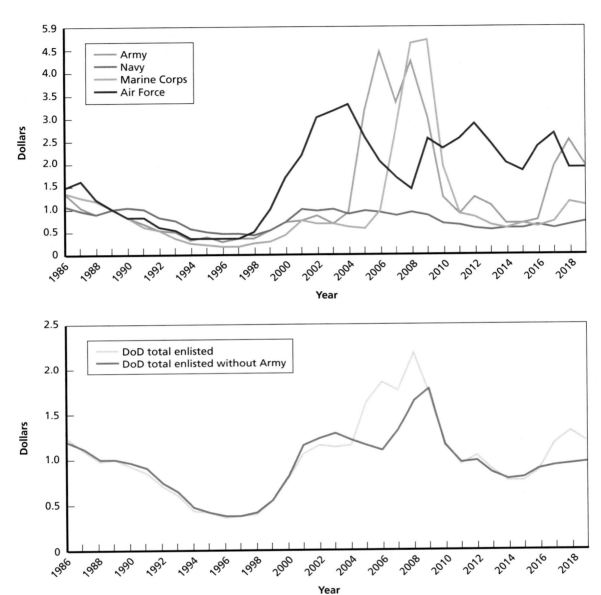

SOURCES: OUSD(P&R) (2018) for 1986–2015; Department of the Air Force, Fiscal Year 2018–2020 Budget Estimates, Military Personnel Appropriation; Department of the Army, Fiscal Year (FY) 2018–2020 Budget Estimates, Military Personnel, Army Justification Book; Department of the Navy, Fiscal Year (FY) 2018–2020 Budget Estimates, Justification of Estimates, Military Personnel, Marine Corps; and Department of the Navy, Fiscal Year (FY) 2018–2020 Budget Estimates, Justification of Estimates, Military Personnel, Navy.
NOTE: The year 1989 = 1.00.

CHAPTER THREE

Comparisons of the Level of Military and Civilian Pay

In this chapter, we assess how RMC has changed through time and how it compares with the wages of similarly educated civilians. As noted in Chapter One, RMC includes basic pay, BAH, BAS, and the federal tax advantage resulting from the allowances not being taxed.[1] The analysis allows us to document when military pay was thought to be adequate in attracting quality recruits and when it was thought to be lacking. We also examine the association between the RMC percentile of the civilian wage distribution and recruit quality to determine the level of military pay necessary to achieve DoD goals.

In this chapter, we present three sets of analyses:

1. trends in RMC percentiles relative to the pay of similar civilians, averaged across all active duty personnel
2. trends in RMC percentiles for specific subgroups of military personnel
3. correlations between metrics of recruit quality and RMC percentiles.

The first analyses provide an overall picture of how RMC compares with civilian pay and how that comparison has changed over time. We conduct this analysis in two ways. First, we compute average RMC percentiles annually from 1980 through 2018, which allows us to consider RMC percentiles in the mid- and late 1980s and in the mid-1990s, the periods that influenced the setting of the 70th percentile as a military pay level benchmark. However, a drawback of the administrative data we use for these analyses is that information on educational attainment appears unreliable. In particular, to assess comparability between military and civilian pay, we require information on the educational attainment of military personnel. But the coding of education categories in DMDC's Active Duty Master Files (ADMF), the administrative data we use, does not appear to be consistent through time. Consequently, as sensitivity analyses, we also compute trends in RMC percentiles averaged across the force using information on educational attainment taken from the Status of Forces Surveys (SOFS)

---

[1]   Note that the definition of RMC was broadened in 1980 to include the variable housing allowance (VHA) and overseas or station housing allowance (SHA) in addition to the basic allowance for quarters (BAQ). Because not all members of the military received these payments, a new measure, "basic military compensation" (BMC) was created to capture all those elements of military compensation that were received by every member of the armed forces. BMC was thus the same as the pre-1980 definition of RMC, in that it included all elements of RMC except for the VHA and SHA. In 1998, the BAH consolidated BAQ with VHA and the overseas housing allowance (OHA) consolidated BAQ with SHA for those living outside the continental U.S., Alaska, and Hawaii. This eliminated the BAQ which had been common to all members regardless of where they lived. Thus, the definition of RMC in our data is slightly different from 1980–1997 and 1998–2018: in the earlier period it includes SHA but in the later period it includes only BAH and not OHA, so it is only applicable to those living in the continental United States, Alaska, and Hawaii. See USD(P&R) (2018).

conducted by DoD's Office of People Analytics. A disadvantage of the SOFS is that the data only go back to the early 2000s and so do not allow us to compute RMC percentiles during the 1980s or 1990s. We find that the estimates of the RMC percentiles differ depending on the source of the data on education.[2] As we argue later in this chapter, future computations of average RMC percentiles should rely on the SOFS for information on the educational distributions of military personnel because these data appear more consistently reported.[3]

We note that neither of the analyses—with education measured by the ADMF or by the SOFS—attempts to incorporate an assessment of whether the educational attainment of military personnel is required for their military duties. To the extent that military personnel seek additional education to improve their post-service military earnings rather than to meet requirements for conducting their military duties, the observed educational attainment of personnel could exceed the military's educational requirements. If so, RMC percentiles will be understated. For example, Smith, Asch, and Mattock (2020) compare enlisted RMC for senior personnel with the civilian earnings of those with similar experience but with an associate's degree versus those with a bachelor's degree. In the former case, the RMC percentile is higher than in the latter case. The implication is that if we generated RMC percentiles that strictly reflected DoD educational requirements, to the extent that these are known, they would be higher than the percentiles we estimate here.

Comparisons of overall trends in the first analyses provide a big picture, but they mask trends for specific subgroups. Thus, the second analyses show trends for specific subgroups. Another advantage of our analyses of specific subgroups is that they do not rely on estimation of the distribution of education of military personnel, as we describe in more detail below.

Finally, we look at the association between measures of recruit quality and weighted RMC through time. These analyses help to underscore the point that RMC should be set at the level necessary to attract the quality of recruits necessary for the DoD to successfully carry out its mission.

Data and Methodology

Because the methodology and data for the analyses presented in this chapter are similar to two other studies that computed RMC percentiles, Hosek et al. (2018) and Smith, Asch, and Mattock (2020), we refer readers to those earlier papers. Here, we provide an overview and only

---

[2]   Our earlier studies, Hosek et al. (2018) and Smith, Asch, and Mattock (2020), used data on the educational distribution of military personnel from the SOFS rather than the ADMF since those studies were concerned with recent comparisons of military and civilian pay.

[3]   In Appendix B, we further discuss how RMC percentiles are sensitive to the choice of education and the civilian comparison group by fixing grade and years of service and comparing service members with similarly educated civilians with the same estimated years of experience.

Our computation of the RMC percentile is weighted to reflect the gender, education, and YOS demographics of the enlisted force and the officer force, thereby providing an overall summary measure of the RMC percentile for enlisted personnel and officers. Because it's a weighted average, changes in military demographics will affect the weighted average, but in a way that is appropriate to decisionmaking. In particular, if the force becomes more senior and better educated, the weighted average will favor the civilian pay of more senior and educated personnel. That said, the weighted average does not provide information to guide targeted pay raises—say, to specific seniority groups. In a companion report (Smith, Asch, and Mattock, 2020) and in previous reports including the 9th and 11th QRMC, RMC percentiles are presented by YOS, thereby providing information on RMC percentiles holding seniority level constant.

discuss in detail issues specific to the current study. This section describes the data used and provides an overview of the process to calculate weighted RMC. The following two sections describe pieces of the calculations which deserve special emphasis: the derivation of education weights for military personnel and the estimation of years of service/years of experience. We then describe the results of the overall weighted RMC calculations.

To make the pay comparisons for the overall weighted RMC, we used data on RMC from DoD's *Selected Military Compensation Tables* (OUSD[P&R], Directorate of Compensation, 1980–2018), also known as the Greenbook. In it, RMC is an average across pay grade and dependency status at each YOS. For the age category analyses, we used data from Active Duty Pay Files provided by DMDC. We use the pay files rather than the Greenbooks because the Greenbooks do not provide information on RMC for subgroups defined by service and other characteristics.

For civilian pay in both of these analyses, we used data on weekly wages and characteristics for civilians from CPS Annual Social and Economic Supplement, also known as the March CPS (U.S. Census Bureau and U.S. Bureau of Labor Statistics, 1980–2019). The CPS, administered by the BLS, uses a representative random sample of the population. As in the 11th QRMC, we used data on full-time, full-year workers, defined as those with a usual workweek of more than 35 hours and who worked more than 35 weeks in the year. All dollar amounts for both military and civilian pay are inflated to 2019 dollars using the Bureau of Labor Statistics Consumer Price Index for All Urban Consumers (CPI-U; U.S. Census Bureau and Bureau of Labor Statistics, 2020). Using these data, we compute civilian pay by gender, education level, and years of labor-force experience.

Computing the overall weighted average RMC in a given year for the first analysis involved three steps. Additional details are provided below.

- **Computed RMC percentiles by YOS and education level for each year:** First, we applied military gender mix weights and computed the civilian wage distribution for each level of education within each year of labor-market experience. Next, treating RMC at each YOS in the military as though it were a wage, we found its placement in the civilian wage distribution for each education level within that year of labor-market experience for each year, i.e., we determined its percentile.[4]
- **Computed the weighted average of RMC percentiles at each YOS:** We estimated the percentage of enlisted members or officers in each education category for each YOS using either the ADMF or the SOFS. Weighting the RMC percentiles by the percentage of individuals in each education category in a given YOS, we then computed the average RMC percentile by YOS for the given year.
- **Computed the weighted average of RMC percentiles across YOS:** We used the number of personnel by YOS to compute an overall weighted average of the RMC percentile for the first twenty YOS by year.

These computations require information on the gender mix of military personnel, the distribution of their educational attainment, and the years of service of military personnel and years of experience of civilian workers. For gender mix, we weight civilian-wage data by the

---

4 For example, as we show in Table 2.4 for enlisted personnel in Smith, Asch, and Mattock (2020), we compute RMC percentiles at each education level for those with 1 YOS, 2 YOS, and so forth.

percentages of men and women in the military in each year using data from *Population Representation in the Military Services: Fiscal Year 2018* ("Pop Rep"; OUSD[P&R], 2020). These percentages vary by year, with the percentage of women generally increasing through time. For example, in 1980, 8.5 percent of enlisted were female, and in 2018 over 16 percent were. Female officers went from 8.2 percent in 1980 to 18.73 percent in 2018. Because we use different sources of data on educational attainment, we discuss that topic in detail separately in the next subsection. We also describe how we approximate years of experience for civilian workers in a subsequent subsection.

Educational Attainment

Over time, the educational attainment of military personnel has increased, such that those in higher grades have reached higher levels of educational attainment than they did in the past (Smith, Asch, and Mattock, 2020). These changes alter the mix of nonmilitary jobs that they can get. For this reason, it is important to compare military RMC with the pay of civilians with more years of formal education as enlisted members progress through their careers. In our by-age-group analyses, we compare enlisted to civilians with a high school diploma and to civilians with some college. Officers are compared separately to both civilians with a bachelor's degree and those with a master's degree or more.

To explicitly take into account changes in education levels through time and during the course of a military career, we constructed a measure of weighted average RMC. Our construction of a weighted average RMC uses estimates of educational attainment by YOS for enlisted and for officers in YOS 0–20. We use two data sources on educational attainment for the purpose of weighting.

The first source is the DMDC ADMF from 1980 to 2018. We used these data to estimate the education distribution by YOS for both enlisted and officers from 1980 through 2018, dropping observations that had missing values or were listed as "unknown" for either education or age.[5] With these data, we consider five levels of education:

1. less than a high school diploma
2. high school diploma
3. some college (including an associate's degree)[6]
4. bachelor's degree
5. master's degree or higher.

In practice, very few officers have less than a bachelor's degree. Table 3.1 shows the education distribution for E-5s and O-6s from the ADMF for 1980, 1990, 2000, 2002, 2010, and 2018.[7] As evident in the table, the coding of education categories does not appear to be con-

---

[5]   Total numbers of individuals in our files, which were provided by DMDC, were broadly similar to numbers provided in the Pop Rep (OUSD[P&R], 2020), although some years matched more closely than others. In our data, 1995–1999 had an unusually high percent of observations with missing values for age. Although we do not use age in the main analyses, we do use it in a robustness check in which we calculate YOS based on age and education. For individuals in 1995–1999 who were also in the 1994 file, we estimated their ages for these years based on their age in 1994 and included them in the sample.

[6]   Those with an "AA, prof. nursing diploma, or 3–4 years of college but no degree" were not broken out in the data until 1995, so we grouped these individuals together with those listed as having "some college" for all years.

[7]   These tabulations are for illustration purposes, to demonstrate differences between the ADMF and SOFS. Because

Table 3.1
ADMF Education Distribution for E-5 and O-3, 1980, 1990, 2000, 2002, 2010, 2018 (%)

Education Level	1980	1990	2000	2002	2010	2018
Enlisted: E-5						
Less than high school diploma	4.51	7.04	2.85	3.77	5.37	2.70
High school diploma	82.23	86.22	63.85	86.99	80.46	72.13
Some college	11.75	4.05	30.61	6.56	10.61	17.35
Bachelor's degree	1.38	2.57	2.56	2.52	3.31	7.18
Master's or more	0.13	0.12	0.14	0.16	0.25	0.63
Officers: O-3						
Less than high school diploma	0.01	0.17	0.08	0.05	0.04	0.06
High school diploma	1.13	0.92	0.46	0.64	0.29	0.71
Some college	3.18	0.69	2.44	3.66	0.67	0.54
Bachelor's degree	62.29	68.27	66.76	70.56	67.85	62.64
Master's or more	33.39	29.96	30.26	25.10	31.15	36.05

SOURCES: ADMF from DMDC, 1980–2018. Excludes those with missing education.

sistent through time, especially for the "some college" category among enlisted. For example, in 1990, the ADMF indicated that 4.05 percent of enlisted E-5s had some college; that figure rose to 30.61 percent in 2000 and dropped to 6.56 percent in 2002.

Although the ADMF data offer many years of estimates for historical comparisons, it is possible that the individual level data may not be updated as military members attain additional education levels after accession. For example, members may not have an incentive to report their additional education if their promotion or other pay or benefits do not depend on it.[8] That is, many military members increase their educational attainment while in service, and the ADMF may underestimate the actual educational attainment of enlisted and officers. This increase in educational attainment changes the RMC percentiles over time, since military members who are more educated have additional outside job opportunities and we must compare them with more educated civilians.

Therefore, we also provide estimates of average RMC using survey data on educational attainment from the SOFS. Although survey data are generally not as accurate or comprehensive as administrative data, we believe that in this case they are more likely to reflect the actual educational attainment of military members than the DMDC ADMF. The data we received are from SOFS from July 2002, August 2004, August 2006, August 2008, June 2010, June 2012, 2014, 2016, 2017, and 2018 (DoD, Office of People Analytics, 2012–2018). A drawback

---

they represent only E-5s and O-3, they do not reflect the education distribution of the entire force.

8   Some education may provide evidence of newly acquired skills and may be relevant to the member's current or anticipated future military duties. As an investment in their future in the military, we would expect this type of education to be more likely to be updated in the ADMF. Conversely, education that is sought to prepare for a post-military career or for personal consumption may be less likely to be reported. As noted earlier, we do not provide any assessment of whether the education members pursue is required for their military assignments.

of these data is that they do not go back far enough in time to allow us to use survey data to compute RMC percentiles during the 1980s or 1990s.

In the SOFS data that we obtained from the Office of People Analytics, enlisted personnel and officers are divided by grade into seven education categories: "non–high school graduate," "high school graduate," "less than one year of college," "one or more years of college but no degree," "associate's degree," "bachelor's degree," and "master's degree or higher." For our analysis of enlisted personnel, we dropped the non–high school graduate, since less than 2 percent of any rank in any year fell in this category, and we consolidated into a "some college" category the categories "less than one year of college," "one or more years of college but no degree," and "associate's degree." This left us with the following categories for enlisted personnel:

1. high school graduate
2. some college
3. bachelor's degree
4. master's degree or higher

For officers, at most 3 percent of officers in a given rank have educational attainment in the first two categories listed above for enlisted, so we excluded them. Consequently, we only use the last two categories. Note, however, that in the officer data, those with an associate's degree are included in the "college graduate" category, and we cannot separate them out. This is in contrast to our enlisted data and the CPS data, where those with an associate's degree are included in the "some college" category.[9]

Table 3.2 shows the education distribution for E-5s and O-3s from the SOFS for 2002, 2010, and 2018. In general, these data show higher education levels than do the ADMF data. For enlisted personnel, this may be due to more consistent coding of education received for those that have more than a high school diploma but less than a bachelor's degree (some college). The survey data may also better reflect education that was attained after members entered the military and progress in their careers.

Years of Service and Years of Experience

The Greenbook provides RMC by YOS, but the CPS does not have data on civilian years of labor-force experience.[10] To compare military and civilian wages adjusted for experience, we

---

[9] In the DMDC ADMF data, where we can separately identify officers with an associate's degree, we see that fewer than 1.1 percent are listed as having associate's degrees in any year since 2000, so we believe that the potential bias caused by including those with associate's degrees in the "bachelor's degree" category is small.

[10] YOS categories in the Greenbooks changed over time. From 1980 to 1992, categories included "Under 2," "2," "3," "4," and then every other year through "22" and then "26." The column for 4 YOS included those with both 4 and 5 YOS, and the last category (26) included all those with greater than that number of YOS. The National Defense Authorization Act for Fiscal Year 1993 effectively established a new longevity step of "over 24" YOS for members of the armed services in pay grades O-6, W-5, W-4, E-9, E-8, and E-7. Thus, in 1993, "24" was added to the pay tables. Similarly, in 2007 the tables were extended to include YOS 28–40 in two-year intervals. In the comparisons with the CPS data, we use our estimates for years of labor force experience and include two years for every year after YOS 4. For example, in the YOS 4 category we include those with 4 and 5 years of estimated experience. Unlike on the military side, we do not include all of those with greater than the maximum year (26 up through 2006 and 40 after) in the last category. Thus, for YOS 22 from 1980 to 1992, for YOS 26 from 1980 to 2006, and for YOS 40 from 2007 to 2018 our samples on the civilian side are constructed with only the given year and the next YOS, whereas the data from the Greenbook include additional YOS.

Table 3.2
SOFS Education Distribution for E-5 and O-3, 2002, 2010, 2018

Education Level	2002	2010	2018
Enlisted: E-5			
Less than high school diploma	1	0	0
High school diploma	21	25	26
Some college	74	67	64
Bachelor's degree	4	7	9
Master's or more	1	1	1
Officers: O-3			
Less than high school diploma	0	0	0
High school diploma	0	0	0
Some college	1	1	1
Bachelor's degree	65	59	59
Master's or more	34	39	40

SOURCES: SOFS, 2002, 2010, 2018.

NOTE: The "bachelor's degree" category for O-3s includes also those with associate degrees as the data that we have for officers records "college graduate or more" and does not distinguish separately between associate's degree holders and bachelor's degree holders. As noted in the text, the percentage of officers with an associate's degree but no bachelor's degree is relatively small.

used assumptions to map age and years of education to years of labor force experience for civilians. Specifically, for high school graduates we subtracted 18 from the person's age in years, for those with some college and associate's degrees we subtracted 20, for college graduates we subtracted 22, and for those with advanced degrees we subtracted 24. For those who started school at a later age or who interrupted their schooling for any reason, these assumptions overstate their experience.

Note, however, that most students initially enrolling in two- and four-year institutions are 19 years or younger (National Center for Education Statistics, 2017). Because we are treating "some college" as two years, we may also be underestimating work experience for some individuals. Conversely, for those who start school late, take a gap year, complete extended religious mission service, or take more than four years for college or more than two years for graduate school, we are assigning them more experience than they have.

A further complication is that years of service on the military side, as recorded in the Greenbooks, may not be the same as total years of experience. For example, if an enlisted high school graduate joined the military at age 22, after two years they would be listed as having two YOS and, in our analysis, would be compared to a 20 year-old civilian high school graduate who we assign two years of experience based on age and education level.[11] For this reason,

---

[11] For our weighted average RMC calculations using education from the ADMF, we computed versions using the YOS as stated in the ADMF and also YOS estimated using the methodology that we use for the CPS. Results were very similar

we also present comparisons by age group later in this chapter. These estimates show similar patterns to the weighted average measures in this section but slightly lower RMC percentiles.

Weighted Average of Regular Military Compensation Percentiles

In this section we consider how RMC compares with the pay of similarly educated civilians over a career and overall, averaged across all personnel. For these analyses, we use RMC from the Greenbooks for 1980 through 2018.

Using the education distribution from the ADMF, we computed the overall weighted average RMC for YOS 1 through 20 for each year from 1980 through 2018 based on the number of personnel by YOS, as listed in the Greenbooks. Estimates are shown in Figure 3.1

Figure 3.1
Enlisted Regular Military Compensation as a Percentile of Civilian Wages Weighted by Level of Education and Year of Service, with Education Estimated from the ADMF, 1980–2018

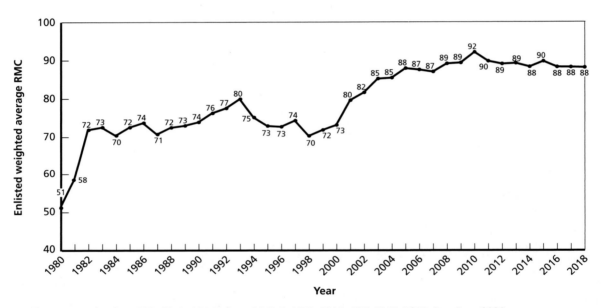

SOURCES: Greenbooks, 1980–2018; ADMF from DMDC, 1980–2018; CPS, 1981–2019; Pop Rep, 2020.
NOTES: We computed the RMC percentile at each level of education, by YOS, as average RMC relative to the civilian wages of full-time, full-year male and female workers, weighted by their proportion in the military. We computed average RMC from the Greenbooks with weights based on the fraction of personnel count, by YOS, which comes from Table A6 in the Greenbooks. Weighted average RMC percentile at each year of service is the sum of the product of the RMC percentile at a given level of education and the fraction of personnel with that level of education. We estimated the education fractions using data from the ADMF from DMDC using all education categories: "dropout," "high school graduate," "some college," "associate's degree," "bachelor's degree," "master's or higher degree." Those with an associate's degree were combined with those with some college for comparison with the civilian data. The overall RMC percentiles shown are the YOS 0–20 weighted average of the average RMC percentile at each YOS, with weights based on the fraction of personnel count by YOS.

between the two approaches, with estimates never differing more than 1 percentile in any given year.

for enlisted personnel and Figure 3.2 for officers. Compared with the wages of similarly edu-cated civilians, enlisted RMC increased sharply in the early 1980s, going from about the 51st percentile in 1980 to the 72nd percentile in 1982. It then stayed relatively stable until the early 1990s, when it began to increase again, peaking at around the 80th percentile in 1993. Through the mid-1990s, the RMC hovered around the 74th percentile of the civilian wage distribution, and it reached a low of about 70 in 1998. Enlisted RMC then began a steady climb and reached the 92nd percentile of the civilian wage distribution in 2010. It has stayed relatively constant at around the 88th percentile since 2011.

Weighted officer RMC also increased sharply in the early 1980s, rising from the 67th percentile of civilian wages in 1980 to the 79th percentile in 1982. It slowly declined through the rest of the 1980s, settling around the 70th percentile in 1987, where it remained until 1998. It then declined further, hitting a low point at the 66th percentile in 2000. Officer RMC increased steadily through the 2000s reaching the 81st percentile in 2009. It has stayed just above or just below the 80th percentile since that time.

Though our methodology differs from Asch, Hosek, and Warner (2001) and Hosek and Sharp (2001), like those studies we find that the RMC percentiles were at around the 70th

Figure 3.2
Officer Regular Military Compensation as a Percentile of Civilian Wages Weighted by Level of Education and Year of Service, with Education Estimated from the ADMF, 1980–2018

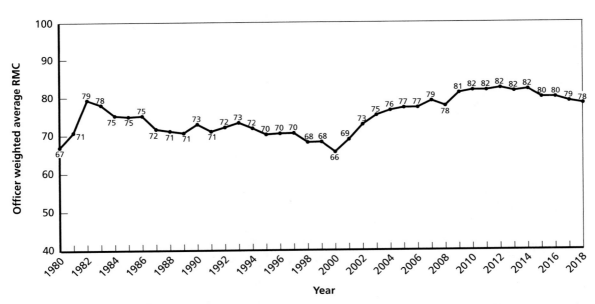

SOURCES: Greenbooks, 1980–2018; ADMF from DMDC, 1980–2018; CPS, 1981–2019; Pop Rep, 2020.
NOTES: We computed the RMC percentile at each level of education, by YOS, as average RMC relative to the civilian wages of full-time, full-year male and female workers, weighted by their proportion in the military. We computed average RMC from the Greenbooks with weights based on the fraction of personnel count, by YOS, which comes from Table A6 in the Greenbooks. Weighted average RMC percentile at each year of service is the sum of the product of the RMC percentile at a given level of education and the fraction of personnel with that level of education. We estimated the education fractions using data from the ADMF from DMDC using all education categories: "dropout," "high school graduate," "some college," "associate's degree," "bachelor's degree," "master's or higher degree." Those with an associate's degree were combined with those with some college for comparison with the civilian data. The overall RMC percentiles shown are the YOS 0–20 weighted average of the average RMC percentile at each YOS, with weights based on the fraction of personnel count by YOS.

percentile in the late 1980s and mid-1990s. The percentiles were actually a bit above the 70th percentile for enlisted personnel and a bit below for officers. RMC rose dramatically in the first half of the 2000s because of pay raises that exceeded the ECI. Since 2010, when pay raises were no longer larger than the change in the ECI, RMC has been relative stable for enlisted personnel, as has the RMC percentile, at around the 89th percentile. Officer RMC has declined somewhat since 2010, to the 78th percentile, still higher than the 70th percentile.

We also estimated weighted average RMC for the first 20 YOS using education distributions from the SOFS. Unfortunately, data from the SOFS does not extend back before 2002. Another drawback of the data we received from the Office of People Analytics is that they do not include information on the education distribution of personnel by YOS. This is in contrast to the ADMF. Thus, for the SOFS data, we needed to impute the education distribution at each YOS from information we received on the education distribution by rank.

We did this imputation in several steps. First, we obtained the joint distribution of personnel by pay grade and YOS from the Greenbooks. This allowed us to compute the percentage of personnel at each pay grade, by YOS. Second, we used these percentages to obtain a weighted average of the education distribution at each YOS (i.e., the percentage with high school, some college, bachelor's degrees, and master's degrees or higher). Third, for each level of education (e.g., high school, some college), we fitted a polynomial curve to its percentages by YOS and then used the fitted curves to predict the percentage, in effect smoothing the percentages.[12] The set of curves for the different levels of education gave us the predicted education distribution by YOS. As with the analysis that computed the education distribution using the ADMF, to compute percentiles, civilian pay by formal education level and age was drawn from the CPS, and military pay for each year of service was drawn from the Greenbooks. Using the RMC percentiles by YOS and education and the distribution of education by YOS, we estimated the average RMC percentile by YOS. Finally, we used the number of personnel by YOS from the Greenbooks to compute an overall weighted average of the RMC percentile for the first 20 YOS.

As can be seen in Figure 3.3, weighted average RMC for enlisted was in about the 72nd percentile in 2002. It then increased, reaching about the 86th percentile in 2010. Since that time, it has remained constant at around the 85th percentile. The figures for 2016–2017 are consistent with the findings in Smith, Asch, and Mattock (2020), since that study also relied on the SOFS. While the trends follow the same pattern as in Figure 3.1, which uses the education distribution from the ADMF, the percentile numbers using the SOFS are lower. Because the SOFSs record higher educational attainment for enlisted and officers than do the ADMF, more weight is placed on RMC percentiles from categories that use more educated civilians as a comparison. More-educated civilians have higher wages, in general, and RMC is a lower percentile of the civilian wage distribution. Consequently, putting more weight on lower percentiles moves the overall weighted average RMC lower.

The same pattern can be seen in Figure 3.4, which shows the weighted average RMC percentile for officers for the first 20 YOS. It is similar to Figure 3.2 but shows slightly lower

---

[12] Specifically, for each education category we regressed the percentage of individuals in that category on a polynomial of YOS and then used the coefficients to get a predicted percentage for each level of education. We normed these percentages to sum to one within a given YOS. For high school and some college, we used a sixth-order polynomial of YOS, for associate's degrees we used a fifth-order polynomial, and for bachelor's and master's degrees we used a fourth-order polynomial. We combine estimates for some college and associate's degrees before norming. Code for reproducing the calculations can be found in Appendix B.

percentiles in some years. Average officer RMC increased gradually from the 70th percentile in 2002 to the 75th percentile in 2008. It then jumped to the 80th percentile in 2010 before declining slightly to the 77th percentile in 2018. Note that because the education distributions for officers are much more similar between the ADMF and the SOFS, with most members having a college degree or a master's degree or higher, the percentile differences between Figures 3.1 and 3.3 are larger than the differences between Figures 3.2 and 3.4.

As noted, although administrative data are generally considered to be more accurate than survey data, we do not believe that is the case in comparing the ADMF and SOFS education data. Education categories do not appear to have been recorded the same through time in the ADMF, and it is likely that these files are not regularly updated when service members achieve higher levels of education as they progress through a military career. For these reasons, we recommend that education data from the SOFS surveys be used in future calculations of the weighted average RMC for both officers and enlisted personnel.

Figure 3.3
Enlisted Regular Military Compensation as a Percentile of Civilian Wages Weighted by Level of Education and Year of Service, with Education Estimated from the SOFSs, 2002–2018

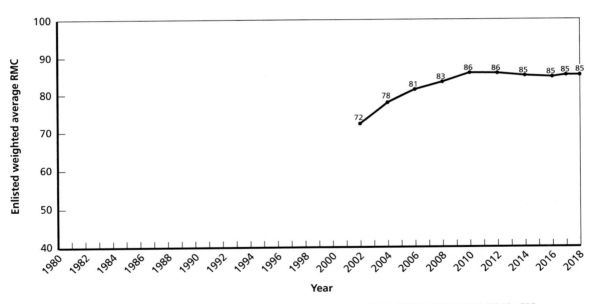

SOURCES: Greenbooks, 1980–2018; SOFS, 2002, 2004, 2006, 2008, 2010, 2012, 2014, 2016, 2017, 2018; CPS, 1981–2019; Pop Rep, 2020.
NOTES: We computed the RMC percentile at each level of education, by YOS, as average RMC relative to the civilian wages of full-time, full-year male and female workers, weighted by their proportion in the military. We computed average RMC from the Greenbooks with weights based on the fraction of personnel count, by YOS, which comes from Table A6 in the Greenbooks. Weighted average RMC percentile at each year of service is the sum of the product of the RMC percentile at a given level of education and the fraction of personnel with that level of education. We estimated the educational attainment distribution and the joint distribution of personnel by pay grade and YOS from the Greenbook and SOFS data. SOFS education categories used included "high school graduate," "less than one year of college," "one or more years of college but no degree," "associate's degree," "bachelor's degree," and "master's degree or higher." The following categories were combined into a "some college" category for comparison with the civilian data: "less than one year of college," "one or more years of college but no degree," "associate's degree." The overall RMC percentiles shown are the YOS 0–20 weighted average of the average RMC percentile at each year of service, with weights based on the fraction of personnel count by YOS.

Figure 3.4
Officer Regular Military Compensation as a Percentile of Civilian Wages Weighted by Level of Education and Year of Service, with Education Estimated from the SOFSs, 2002–2018

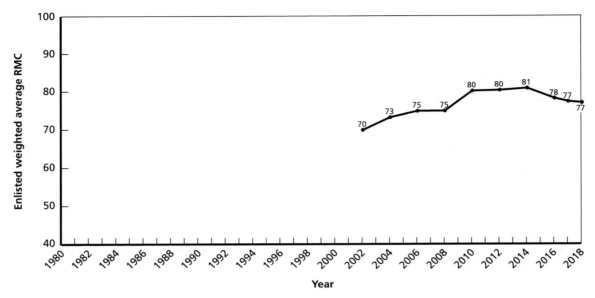

SOURCES: Greenbooks, 1980–2018; SOFS, 2002, 2004, 2006, 2008, 2010, 2012, 2014, 2016, 2017, 2018; CPS, 1981–2019; Pop Rep, 2020.
NOTES: We computed the RMC percentile at each level of education, by YOS, as average RMC relative to the civilian wages of full-time, full-year male and female workers, weighted by their proportion in the military. We computed average RMC from the Greenbooks with weights based on the fraction of personnel count, by YOS, which comes from Table A6 in the Greenbooks. Weighted average RMC percentile at each year of service is the sum of the product of the RMC percentile at a given level of education and the fraction of personnel with that level of education. We estimated the educational attainment distribution and the joint distribution of personnel by pay grade and YOS from the Greenbook and SOFS data. SOFS education categories used included "college graduate or more," and "advanced degree." Note that "college graduate or more" includes those with associate's degrees in the SOFS data but not in the CPS data. The overall RMC percentiles shown are the YOS 0–20 weighted average of the average RMC percentile at each year of service, with weights based on the fraction of personnel count by YOS.

Trends in the Regular Military Compensation Percentile for Selected Age and Education Groups, 1994–2018

We also computed the RMC percentile for 1994 through 2018 for specific groups defined by service, gender, education level, and age. We conducted this analysis for each service, but we present only the results for Army men because results were similar across services. To compute RMC in these analyses, we use cross-section data on males from the given age group and rank (officer or enlisted) from the DMDC Active Duty Pay Files.[13] That is, unlike the computations shown in Figures 3.1–3.4, we do not use the Greenbooks for estimates of RMC. For each subgroup, defined by service, gender and age group, we compare RMC to the pay of civilians of similar gender and age with a pre-defined education attainment. Specifically, for Army men, we make comparisons of individuals in the following groups:

---

[13] Computing RMC with the military-pay files required that we compute the tax advantage. It is based on taxable (basic pay) and nontaxable (BAS and BAH) income, number of dependents, and marital status. Additional details can be found in Hosek et al. (2018, pp. 4–8).

- enlisted members ages 23–27 compared with civilian high school graduates in this age range
- enlisted members ages 28–32 compared with civilians with some college in this age range
- officers ages 28–32 compared with civilians with bachelor's degrees in this age range
- officers ages 33–37 compared with civilians with master's degrees or higher in this age range.

The implicit assumption is that the relevant civilian opportunity wage for enlisted males ages 23–27 is the civilian pay of male high school graduates of similar ages, whereas the relevant civilian wage for enlisted males ages 28–32 is the civilian pay of males with some college of similar ages. Note that these comparisons differ from the weighted average RMC computations presented above because they do not adjust for YOS and because some individuals enter service at older ages and have fewer YOS than one would expect based on their ages.

Overall, in Figures 3.5–3.8, we find that RMC for these groups increased substantially from 1994 to 2010 and then stayed roughly constant through 2018. The increase was driven by factors discussed in Chapter One including a restructuring of the basic-pay table from 2001 through 2003, higher-than-usual basic-pay increases from FY 2000 to FY 2010, increases in BAH implemented in the first part of the 2000s to cover the full cost of housing, and increases in housing cost that resulted in further BAH increases.

RMC percentiles are also affected by the trends in civilian wages. As shown in the figures for the 50th percentile of civilian earnings, median civilian earnings generally trended upward from the mid-1990s through 2000 across the subgroups. Median earnings then trended downward after 2000, especially during the Great Recession beginning in late 2008 and 2009, leveling off around 2012–2014. Since then, median civilian wages have tended to increase, especially in 2017–2018.

From 2010 onward, the figures indicate that RMC was between

- the 81st and 87th percentiles for enlisted members ages 23–27 compared with civilian high school graduates
- the 73rd and 82nd percentiles for enlisted members ages 28–32 compared with civilians with some college
- the 82nd and 89th percentiles for officers ages 28–32 compared with civilians who were four-year college graduates
- the 68th and 76th percentiles for officers ages 33–37 compared with civilians with master's degrees or higher.

Associations Between Measures of Recruit Quality and Weighted RMC Percentiles

Military compensation is one of the primary tools used by the services to acquire the quantity and quality of personnel they need. In this section, we examine the association between RMC percentile and the quality of recruits over time.[14] Our measure of high quality focuses on those

---

[14] Following previous work, such as Hosek, Asch, and Mattock (2012), Hosek et al. (2018), and Smith, Asch, and Mat-

Figure 3.5
Civilian Wages for High School Graduate Men and Median Regular Military Compensation for Army Enlisted, Ages 23–27, Calendar Years 1994–2018, in 2019 Dollars

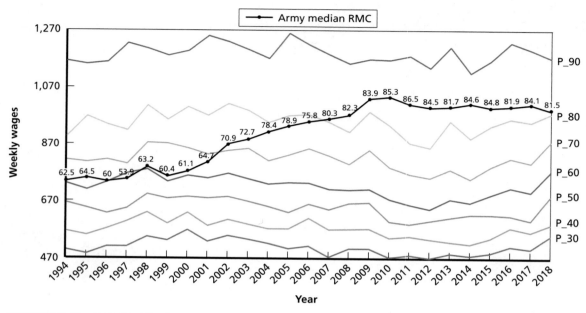

SOURCES: Active Duty Pay Files from DMDC; CPS, 2019.
NOTES: The reference population is men ages 23–27 who reported high school completion as their highest level of education, worked more than 35 weeks in the year, and usually worked more than 35 hours per week. We computed the weekly wage by dividing annual earnings by annual weeks worked. The colored lines depict the wages at the indicated percentiles (on the right axis) of the wage distribution for this population. For instance, at the 70th percentiles (denoted P_70), 30 percent of the population had higher wages and 70 percent had lower wages. The black line depicts median RMC for Army enlisted between ages 23 and 27. The numbers above the RMC line are the percentiles at which RMC stood in the population's wage distribution.

recruits who have no prior service, that is, non–prior service accessions. Past research has found a positive relationship between military compensation and recruit quality.[15] Our focus here is not on replicating this past work but on examining how recruit quality changed as RMC percentiles changed, with a particular emphasis on how recruit quality changed in recent years, when RMC has exceeded the 70th percentile of earnings of similar civilians.

Before presenting the results, it is important to reiterate a point discussed in Chapter Two: Raw trends in the quality of recruits over time do not account for other factors that were also changing over this time period, such as the outside job options, recruiting goals, or the perceived risks of a military career. While previous research has found that there is a positive relationship between relative military pay and high quality even after controlling for many other factors that influence recruit quality,[16] we note that the relationships between recruit

---

tock (2020), our analysis focuses on the relationship between RMC and recruiting. We could have done a similar analysis with retention, but we chose to focus on recruiting because our retention metrics include individuals who are not necessarily free to make a decision because they are still under a service obligation. Furthermore, because the military permits virtually no lateral entry, recruit quality is important in terms of determining the quality of the overall force, as discussed in Asch, Romley, and Totten (2005).

[15] See Asch (2019a) for a summary of these studies.

[16] There is a long literature exploring the relationship between recruit quality and RMC, controlling for other factors.

Figure 3.6
Civilian Wages for Men with Some College and Median Regular Military Compensation for Army Enlisted, Ages 28–32, Calendar Years 1994–2018, in 2019 Dollars

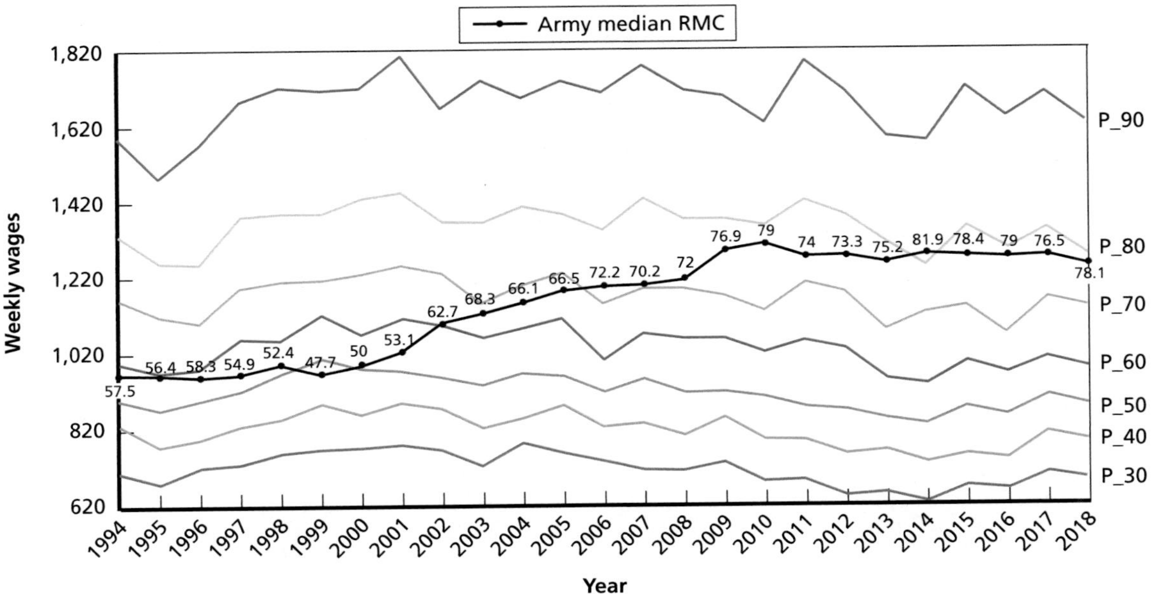

SOURCES: Active Duty Pay Files from DMDC; CPS, 2019.
NOTES: The reference population is men ages 28–32 who reported some college as their highest level of education, worked more than 35 weeks in the year, and usually worked more than 35 hours per week. We computed the weekly wage by dividing annual earnings by annual weeks worked. The colored lines depict the wages at the indicated percentiles (on the right axis) of the wage distribution for this population. For instance, at the 70th percentile, 30 percent of the population had higher wages and 70 percent had lower wages. The black line depicts median RMC for Army enlisted between ages 28 and 32. The numbers above the RMC line are the percentiles at which RMC stood in the population's wage distribution.

quality and weighted average RMC percentile presented here should be viewed as descriptive and do not represent the causal effect of increases in RMC percentile on the quality of military recruits.

Another important point is that this analysis of the association between recruit quality and RMC percentiles is not intended to imply that military pay is the only or even the primary force management policy tool available to the services for influencing recruiting outcomes. Past research has shown that policies such as bonuses, recruiters, and advertising also affect recruiting outcomes and, importantly, are more cost-effective policies than pay as a means of addressing recruiting shortfalls (e.g., Warner, 2010; Simon and Warner, 2007). That said, we focus here on the association between RMC percentile and outcomes because of interest in how recruiting outcomes have changed since the period when the 70th percentile was set.

As shown in Figure 3.9, the percentage of high-quality non–prior service recruits has varied substantially over time, from 35 percent in 1980 to a high of 77 percent in 2011. Per-

In earlier work (Hosek et al., 2018; Smith, Asch, and Mattock, 2020), we reviewed some of this literature and did more explicit modeling of the relationship between recruit quality and the ratio of military and civilian pay, controlling for the unemployment rate, gender, service recruiting goals, deployment, and the Post-9/11 Veterans Educational Assistance Act of 2008. We did not control for enlistment bonuses, number of recruiters, or advertising because these quantities are determined by the services and are endogenous to the model. Unfortunately, we did not have exogenous variation in these variables to separately identify their effects on quality independent of RMC.

Figure 3.7
Civilian Wages for Men with Four-Year College Degrees and Median Regular Military Compensation for Army Officers, Ages 28–32, Calendar Years 1994–2018, in 2019 Dollars

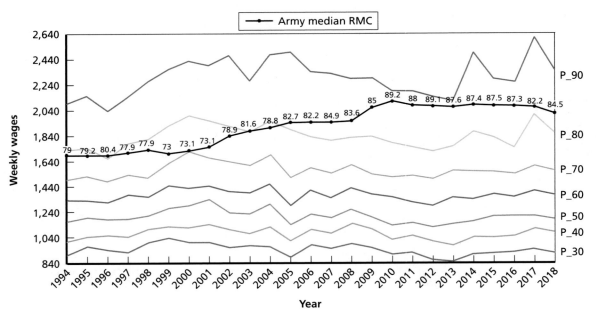

SOURCES: Active Duty Pay Files from DMDC; CPS, 2019.
NOTES: The reference population is men ages 28–32 who reported bachelor's degrees as their highest level of education, worked more than 35 weeks in the year, and usually worked more than 35 hours per week. We computed the weekly wage by dividing annual earnings by annual weeks worked. The colored lines depict the wages at the indicated percentiles (on the right axis) of the wage distribution for this population. For instance, at the 70th percentile, 30 percent of the population had higher wages and 70 percent had lower wages. The black line depicts median RMC for Army officers between ages 28 and 32. The numbers above the RMC line are the percentiles at which RMC stood in the population's wage distribution.

centage high-quality generally moved together with weighted average RMC percentiles from 1980 through 1998 and again from 2011 to 2017.[17] However, the two series diverged between 1998 and 2011, and percentage high-quality fell sharply from 2004 to 2008 before rising sharply from 2008 to 2011. While part of this decline in quality was likely influenced by favorable economic conditions prior to 2008 and then the Great Recession starting in that year, the RMC percentile of the civilian wage distribution continued to increase throughout this period. The decrease in recruit quality was also only evident in the Army, making it unlikely that general economic conditions can explain all of the divergence between RMC percentile and recruit quality. As detailed in Hosek et al. (2018) and as discussed in Chapter Two, non–prior service recruit quality increased between 2000 and 2017 for the Air Force, Navy, and Marine Corps but not the Army. The Air Force, Navy, and Marine Corps increased their percentages of accessions who were high-quality and had, or reached, a very high percentage of accessions who were non–prior service Tier 1 recruits. The Army's percentage of accessions who were high-quality fell after 2004, then rebounded to its initial level by 2010, and then stayed there. Its Tier 1 percentage bottomed out in 2007 and then rose to a stable level closer to that of the other services by 2010. The Army's percentage of accessions in AFQT categories I–IIIA in the active component declined fairly steadily after 2004 but showed an uptick in 2018. Over the

---

[17] Weighted average RMC in this graph is the same as that shown in Figure 3.1.

Figure 3.8
Civilian Wages for Men with Master's Degrees or Higher and Median Regular Military Compensation for Army Officers, Ages 33–37, Calendar Years 1994–2018, in 2019 Dollars

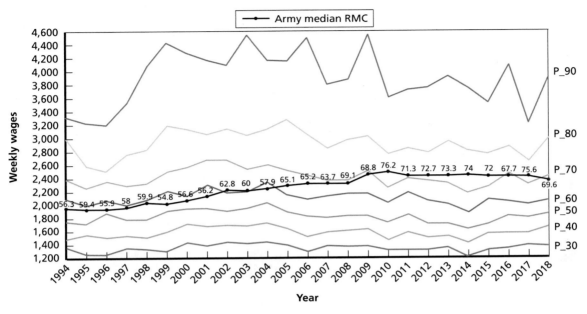

SOURCES: Active-duty pay files from DMDC; CPS, 2019.
NOTES: The reference population is men ages 33–37 who reported master's degrees or higher as their highest level of education, worked more than 35 weeks in the year, and usually worked more than 35 hours per week. We computed the weekly wage by dividing annual earnings by annual weeks worked. The colored lines depict the wages at the indicated percentiles (on the right axis) of the wage distribution for this population. For instance, at the 70th percentile, 30 percent of the population had higher wages and 70 percent had lower wages. The black line depicts median RMC for Army officers between ages 33 and 37. The numbers above the RMC line are the percentiles at which RMC stood in the population's wage distribution.

same period, this percentage increased in the other services. Hosek et al. (2018) note that the fall in quality among Army accessions was driven by both a reduction in Tier I recruits and a smaller decrease in the percentage of overall recruits who scored in category 3A or above.[18]

Figure 3.10 shows a scatterplot of weighted average enlisted RMC percentile and percentage high-quality recruits for 1985 to 2017. The measure of weighted average RMC percentile shown here is calculated using the education distribution from the ADMF and is the same as that shown in Figure 3.1. Figure 3.10 also shows a best fit line through the data and the slope of the line is positive and statistically significant at less than the 0.01 level, demonstrating a positive association between recruit quality and weighted average RMC percentiles.[19] That said, if the Army is excluded from the computations of high-quality, the relationship is even stronger, 0.9637 for the slope coefficient rather than 0.4278 (Figure 3.11).

---

[18] For additional discussion on why the Army failed to increase its quality at a time of increasing RMC, see Hosek et al. (2018, pp. 71–73).

[19] The best fit line is found by an ordinary least squares regression. The intercept (30.665) and slope (.4278) of the regression are shown in the figure. Because both percentage of high-quality recruits and weighted average RMC percentile were low in 1980–1982, these points are outliers. While they record valid data, they may not be relevant for informing present policy, since both percentage of high-quality recruits and weighted average RMC are much higher today. If we include the years before 1985, the relationship between percentage high-quality recruits and weighted average RMC percentile is still positive and statistically significant, and the magnitude of the association is almost double.

Figure 3.9
Weighted Average Enlisted Regular Military Compensation as a Percentile of Civilian Wages Weighted by Level of Education and Year of Service, with Education Estimated from the ADMF and Percentage High-Quality Recruits, 1980–2017

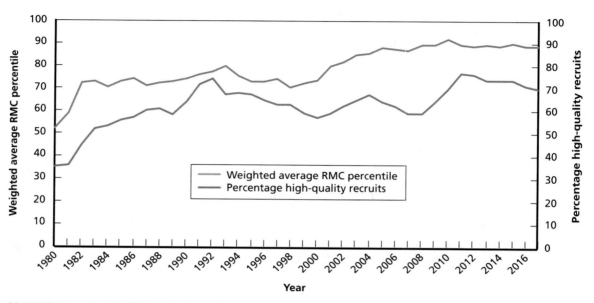

SOURCES: Greenbooks, 1980–2018; ADMF from DMDC, 1980–2018; CPS, 1981–2019; Pop Rep, 2020.
NOTES: We computed the RMC percentile at each level of education, by YOS, as average RMC relative to the civilian wages of full-time, full-year male and female workers, weighted by their proportion in the military. We computed average RMC from the Greenbooks with weights based on the fraction of personnel count, by YOS, which comes from Table A6 in the Greenbooks. Weighted average RMC percentile at each year of service is the sum of the product of the RMC percentile at a given level of education and the fraction of personnel with that level of education. We estimated the education fractions using data from the ADMF from DMDC using all education categories: "dropout," "high school graduate," "some college," "associate's degree," "bachelor's degree," "master's or higher degree." Note that those with an associate's degree were combined with those with some college for comparison with the civilian data. The overall RMC percentiles shown are the YOS 0–20 weighted average of the average RMC percentile at each YOS, with weights based on the fraction of personnel count by YOS. High-quality recruits are those who are Tier 1 and score in the upper half of the AFQT score distribution.

A key question motivating our analysis is whether the RMC percentile needs to be higher today to achieve the quality that was achieved in the 1990s, the period relevant to the setting of the 70th percentile. If it were the case that the RMC percentile needed to be higher to achieve the quality of the 1990s, we would expect to see little change in recruit quality relative to the 1990s as the RMC percentile increased in more recent years—that is, a weak association between RMC percentile and percentage high-quality. Instead, as shown in Figure 3.11 and to a lesser extent in Figure 3.10, where the Army is included, the percentage of recruits that were high-quality increased and did not remain the same as RMC increased—that is, we find a strong positive association between RMC percentile and percentage high-quality. The figure shows that in the late 1980s and mid 1990s, the RMC percentile was generally between the 70th and 75th percentile, and the percentage of recruits that were high-quality was from 60 to 65 percent. But since 2010, the RMC percentile has been between the 80th and 90th percentile, and the percentage high-quality has been between 70 and 85 percent. The implication is that the higher RMC percentiles in recent years sustained higher-quality accession cohorts than in the late 1980s and mid-1990s, when the 70th percentile was set. A similar conclusion can be drawn from the scatter plot in Figure 3.10 that includes the Army. Recruit quality has

Figure 3.10
Association Between Enlisted RMC Percentile and Percentage High-Quality Recruits, All Services, 1985–2017

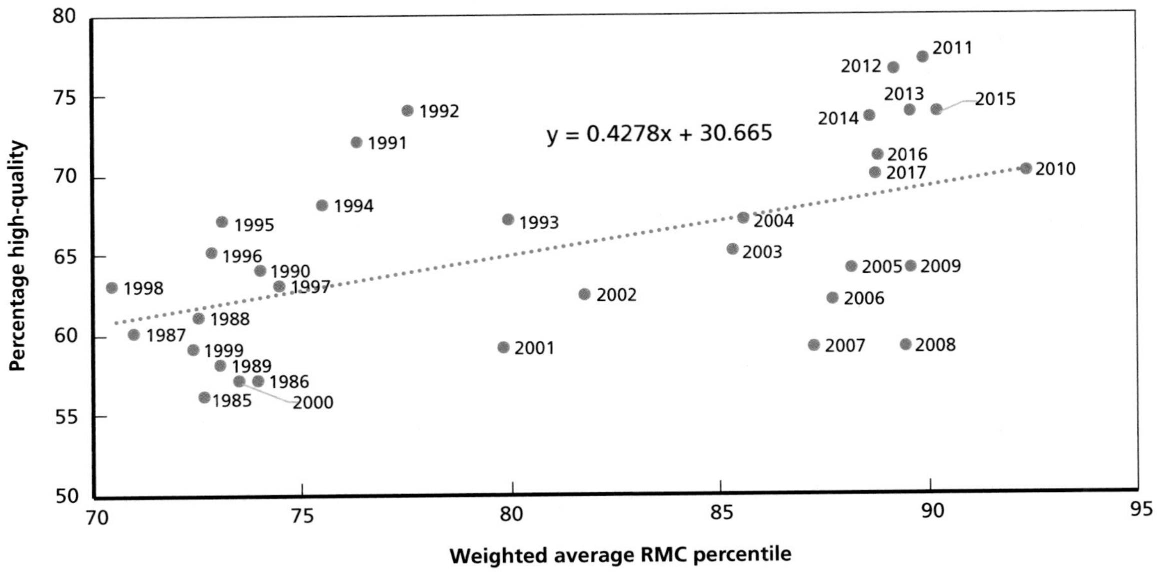

SOURCES: Greenbooks, 1985–2018; ADMF from DMDC, 1985–2018; CPS, 1986–2019; Pop Rep, 2020.
NOTES: We computed the RMC percentile at each level of education, by YOS, as average RMC relative to the civilian wages of full-time, full-year male and female workers, weighted by their proportion in the military. We computed average RMC from the Greenbooks with weights based on the fraction of personnel count, by YOS, which comes from Table A6 in the Greenbooks. Weighted average RMC percentile at each year of service is the sum of the product of the RMC percentile at a given level of education and the fraction of personnel with that level of education. We estimated the education fractions using data from the ADMF from DMDC using all education categories: "dropout," "high school graduate," "some college," "associate's degree," "bachelor's degree," "master's or higher degree." Note that those with an associate's degree were combined with those with some college for comparison with the civilian data. The overall RMC percentiles shown are the YOS 0–20 weighted average of the average RMC percentile at each YOS, with weights based on the fraction of personnel count by YOS. High-quality recruits are those who are Tier 1 and score in the upper half of the AFQT score distribution.

been higher since 2010 than in the late 1980s and mid-1990s, albeit not as high when the Army is excluded, and the RMC percentile has been higher, too.

Summary

We computed weighted RMC percentiles for enlisted personnel and officers, adjusting for the education distribution of military personnel. We find sharp increases in RMC in the early 1980s, gradual and steady increases through 1993, declines from 1994 to 1998, substantial increases between 1998 and 2010, and a leveling off since 2010. This latest period of increase reflects the relatively fast military pay growth from the late 1990s to 2010, as well as a downward trend in real civilian wages. Similar patterns emerge in our by-age-group analyses. We also show that weighted average RMC percentile is strongly correlated with recruit quality through time. Finally, we argue that in the future data sources that more accurately reflect changes in educational attainment among military members, such as the SOFS, should be used when comparing military pay with civilian wages.

Figure 3.11
Association Between Enlisted RMC Percentile and Percentage High-Quality Recruits, All Services Except the Army, 1985–2017

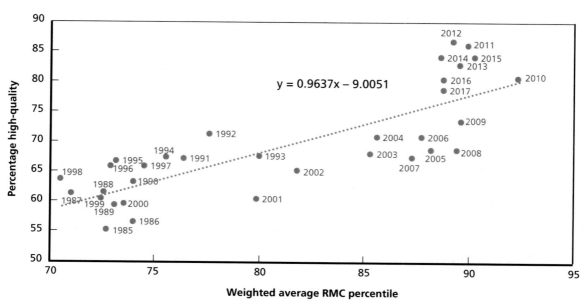

SOURCES: Greenbooks, 1985–2018; ADMF from DMDC, 1985–2018; CPS, 1986–2019; Pop Rep, 2020.
NOTES: We computed the RMC percentile at each level of education, by YOS, as average RMC relative to the civilian wages of full-time, full-year male and female workers, weighted by their proportion in the military. We computed average RMC from the Greenbooks with weights based on the fraction of personnel count, by YOS, which comes from Table A6 in the Greenbooks. Weighted average RMC percentile at each year of service is the sum of the product of the RMC percentile at a given level of education and the fraction of personnel with that level of education. We estimated the education fractions using data from the ADMF from DMDC using all education categories: "dropout," "high school graduate," "some college," "associate's degree," "bachelor's degree," "master's or higher degree." Note that those with an associate's degree were combined with those with some college for comparison with the civilian data. The overall RMC percentiles shown are the YOS 0–20 weighted average of the average RMC percentile at each YOS, with weights based on the fraction of personnel count by YOS. High-quality recruits are those who are Tier 1 and score in the upper half of the AFQT score distribution.

CHAPTER FOUR
An Alternative Index for Setting the Annual Pay Adjustment

The importance of adjusting military pay in an appropriate manner is motivated by two primary concerns. One is that allowing military pay to become uncompetitive with pay in the civilian labor market will lead to a decline in both the quantity and quality of new accessions into military service and retention of experienced personnel. A second concern is that if pay grows more quickly than required for maintaining a capable, fully staffed force, the flexibility of the services to deploy financial resources in other areas, such as hardware and technology, will be reduced. Additionally, paying wages higher than necessary for the national defense represents an inefficient use of taxpayer resources.

Since 2003, the ECI has been the official source of statutory guidance in setting annual military pay increases, but it has served in this role unofficially since 1990.[1] The ECI is a quarterly price index constructed by the BLS to measure the change in the price of total compensation per employee hour worked using survey data collected from a large, representative sample of U.S. employers.[2] However, since the early 1990s, there has been interest among military policymakers in exploring an alternative to the ECI, and the Defense Employment Cost Index (DECI) was considered as a candidate index to fill this role by the 7th QRMC. As mentioned in Chapter One, in 1992 the 7th QRMC chose not to recommend that the DECI replace the ECI but recognized its advantages and recommended that its continued development be supported by DoD.

This chapter presents an update and extension of the DECI, as well as a reassessment of its suitability as a candidate index in guiding the annual military pay increase.[3] The long interim between the 7th and 13th QRMCs allows us the benefit of nearly 30 years of additional data to consider, a period during which the U.S. experienced both dramatic macroeconomic shocks and sustained the extensive deployment of active duty service members in two major regional contingencies. In this chapter, we first reintroduce the DECI and discuss its composition and construction for those not familiar with the original Hosek 1992 report. By way of comparison, we also discuss the composition and construction of the ECI. We then compare the guidance provided by both the ECI and the DECI over a period of more than

[1] Section 602 of the FY2004 NDAA (Pub. L. 108-136, 2003; 117 Stat. 1498, amending 37 USC 1009) formalized the use of the ECI in setting the annual military pay adjustment. The Federal Employees Pay Comparability Act of 1990 (Pub. L. 101-509, Section 529, 1990) set the ECI as the official measure used to set annual increases in the GS scale used for civil servant salaries, and, since 1967, when the annual military pay adjustment process was established, it has been tied to the adjustment of the GS scale.

[2] In 2001, the ECI comprised more than 30,000 occupational observations from approximately 7,000 private firms (Ruser, 2001).

[3] This chapter draws primarily from the 7th QRMC (DoD, 1992), Hosek 1992, Goldich (2005), and Ruser (2001).

three decades and explore the reason these indices differ over this period. Finally, we assess the relevance of the information regarding the path of the annual military pay raise imparted by each index through a set of exercises comparing the ability of the DECI and the ECI to predict changes in measures of accession and continuation over more than three decades.

The Employment Cost Index (ECI)

The underlying data used to construct the ECI comes from the BLS's National Compensation Survey, a large survey of compensation costs among business establishments (and governmental entities). In the ECI, a sample of jobs is chosen from among the employers completing this survey, and then this collection of jobs is held fixed for multiple years (up to five) in order to sample within-job earnings changes.[4] The ECI is a "Laspeyres" index, meaning that it calculates the average change in total labor compensation across multiple years using a set of weights, one for each occupation-by-industry group, that reflect the proportional share of each employment category in the economy during a chosen baseline period. This share is then held fixed over time. Specifically, the index of changes in earnings, e, in year t, relative to a baseline year ($t = 0$), where the earnings weights, w, are fixed, is defined as

$$ECI_t = \left[\frac{\sum_i w_{i0} e_{it}}{\sum_i w_{i0} e_{i0}} \right] \times 100.$$

In the numerator, the product of the average earnings in the ith employment category in period t and the weight assigned to this employment category in the baseline period, t_0, are summed over all employment categories. The denominator is computed similarly, but both the weights *and* the earnings are from the baseline period, t_0. The resulting ratio is multiplied by 100. Thus, only one element—the earnings measure in the numerator—is changing over time, with the other elements held fixed as long as the initial set of weights is in place. In the case of the ECI, the set of occupation-by-industry weights are updated approximately once per decade to reflect periodic changes in the composition of employment in the economy (the ECI currently uses weights calculated in 2012).

The fact that the ECI is representative of the average change in compensation paid in the civilian labor market may be viewed as supportive of the key role it has been given in informing the appropriate change in level of basic pay in the military. But multiple aspects of the ECI may limit its accuracy as a guide to the appropriate change in level of military compensation.

Demographic Differences Between the Military and the Civilian Labor Force

The ECI does not control for the age, educational attainment, or gender composition of workers. This omission does not affect the ability of the ECI to fulfill its primary goal, generating an unbiased estimate of the change in the average price of labor in the economy. However, age, educational attainment, and gender are empirically important predictors of occupation and, by extension, compensation. Thus, these factors are relevant to the civilian earnings opportunities

---

[4]   For a detailed description of how the ECI sample is selected and replaced, see Ruser (2001).

of military personnel. For example, many high-paying white-collar occupations require the completion of a baccalaureate degree; many lower-paying, unskilled jobs are disproportionately held by younger workers; within a given occupation, older, more experienced workers are typically paid more than younger workers with less experience; and many occupations exhibit high (though generally declining) levels of gender segregation (Cortes and Pan, 2017). All these factors, not just the change in average compensation in a given occupation-by-industry category, are important in adjusting military pay in an appropriate manner.

The age, educational attainment, and gender composition of the military differs significantly from that of the civilian labor force. Figures 4.1 and 4.2 compare the age and educational attainment composition of active duty military personnel and the civilian labor force across time. As can be seen in Figure 4.1, around 87 percent of active duty military personnel are below the age of 37, and this share has varied little across multiple decades. In contrast, only around half of the civilian labor force is below age 37 over these same decades, and this proportion has declined from nearly 60 percent to just over 40 percent between 1982 and 2018 because of the aging of the Baby Boomer generation and declines in the fertility rate over the 1960s and 1970s.

Similarly, the composition of educational attainment in the private industry workforce in the United States differs significantly from that of the military. As shown in Figure 4.2, panel A, individuals with a high school diploma as their highest level of educational attainment make up around 77 percent of all active duty service members. While the size of this share has declined somewhat over time in the civilian labor market (panel B),[5] those with a high school diploma alone made up 59 percent of workers in 1982 and just 34 percent by 2018, a decline of over 70 percent.[6] Thus, in both age and education domains, the ECI has become increasingly less representative of the composition of active duty military personnel over time.

While the share of women among active duty military personnel has increased from around 10 percent in the early 1980s to around 16 percent currently, the share of women in the full-time civilian labor force has remained around 42 percent across the past three decades. Additionally, the rate of growth of female earnings has outpaced men over this same time period likely due to both the rising educational attainment of women and related inroads into previously male-dominated occupations (Blau and Kahn, 2017).

Measuring Compensation Using a Job-Based Sample

A second aspect of the ECI that may limit its accuracy as a guide to the appropriate level of military compensation is the approach used for measuring compensation. Using an employer-

---

5   The education composition of the active force is based on tabulations using the ADMF. As discussed in Chapter Three, these data have the advantage of allowing us to observe trends over a long period of time, including periods prior to the 2000s, but may not be entirely accurate, as suggested by data from the SOFS.

6   As we discussed in Chapter Three, the observed educational attainment of military personnel could exceed the military's educational requirements if individuals attain greater education to improve their post-service earnings rather than to meet military requirements for conducting their duties. If so, the DECI computation will give greater weight to better-educated civilians than what would occur if we could incorporate DoD's educational requirements. Because better-educated civilians have experienced faster pay growth (see Figures 4.4 and 4.5), the DECI will appear higher than what we would have computed had we used DoD's educational requirements. If the DECI is used to guide the annual pay adjustment, it would imply faster military pay growth than what would have been suggested by a DECI based on lower educational attainment.

Figure 4.1
Age Composition of Active Duty Military and Civilian Labor Force

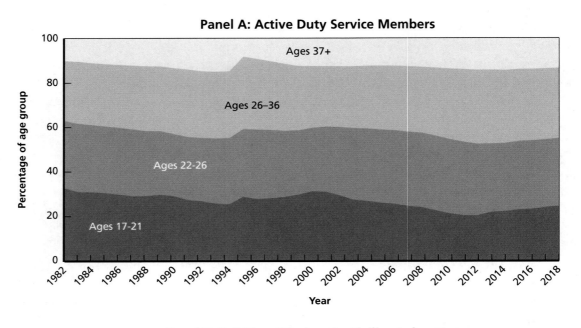

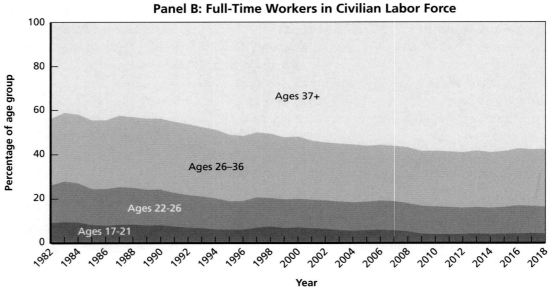

SOURCES: Civilian labor force age composition uses CPS ORG data (April–September, 1982–2019) from the Integrated Public Use Microdata Series (IPUMS) (Flood et al., 2020). Military age composition uses ADMF data from DMDC over the same period.
NOTE: Data points for these and other figures are in Appendix F.

Figure 4.2
Educational Attainment of Active Duty Military and the Civilian Labor Force

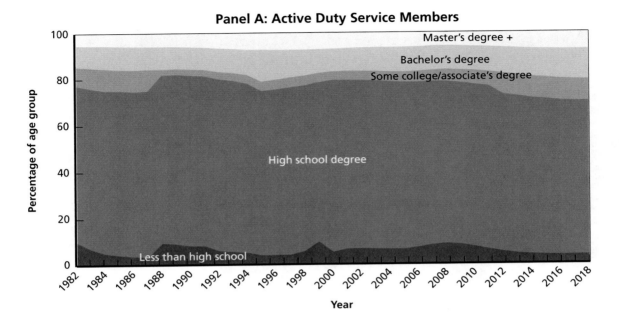

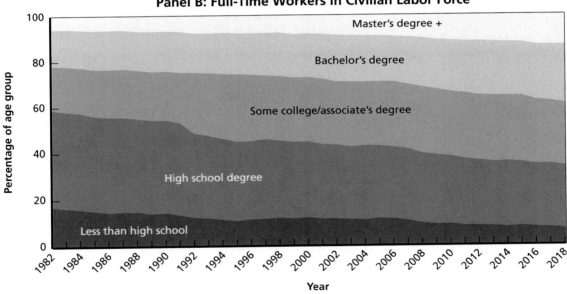

SOURCES: Civilian labor force educational attainment composition uses CPS ORG data (April–September, 1982–2019) from IPUMS (Flood et al., 2020). Military educational attainment composition uses ADMF data from DMDC over the same period.
NOTE: Data points for these and other figures are in Appendix F.

cost-based approach may lead to an earnings index that is less representative of the earnings that would be faced by a worker newly entering the civilian labor market.

Specifically, measuring within-job earnings changes across economic downturns results in a sample that will overrepresent "job-stayers," and the earnings of these workers may significantly differ from the earnings of workers in the overall labor force. Figure 4.3 shows the change in job quits and job hires, in thousands of workers, on the right and left y-axis, respectively, across the period of the Great Recession of 2008–2009. Over these years, the level of both measures of the flow of workers into and out of jobs declined dramatically. When this occurs, job tenure will increase on average, which will lead to the ECI measuring compensation changes for relatively more-experienced workers in recessionary periods. Since the earnings of most jobs exhibit positive returns to experience, this tendency will result in the ECI moderating or reducing the negative effect of an economic downturn on earnings.[7] Furthermore, if firms are disproportionately likely to lay off less experienced workers during a downturn, this will cause the average wages in a given employment sector to rise, further overstating changes in earnings from the perspective of an individual worker. Both of these effects work in the opposite direction of overall compensation changes during both recessions and economic booms as well (though not necessarily in a symmetrical fashion). In the case of expansionary hiring periods, when voluntary job changes are at a higher level, the average experience of a worker in a given job may decline, and the ECI will be less sensitive than the DECI to increasing earnings opportunities in the civilian labor market. For these reasons, the 7th QRMC pointed to the use of worker-based survey data, rather than employer-based survey data, as a virtue of the DECI.[8]

Figure 4.3
Hires and Quits Across Economic Contractions

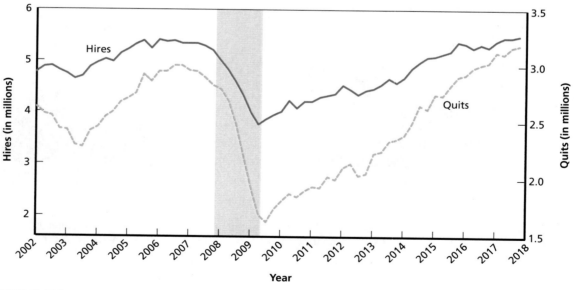

SOURCE: BLS.

---

[7]   For more on the relationship between hires, quits, and the business cycle, see Clark (2004).

[8]   One potential virtue of the ECI is that the earnings data reported in employer surveys are likely to be drawn from

Relatedly, the use of fixed employment weights over a decade may mask significant changes in the composition of jobs in the economy that are relevant for capturing the external civilian earnings opportunities of military personnel. Shifts in employment across sectors often signal important changes in the relative return to different skills over time. The ECI, however, will replace jobs lost through such sectoral changes with the remaining jobs in the same occupation-by-industry group. To illustrate the potential problem with this aspect of the index, consider the example of the remarkable decline in manufacturing jobs in recent decades. One recent study estimated that between 2002 and 2007, U.S. manufacturing jobs declined 18 percent (Pierce and Schott, 2016), yet the occupation-by-industry weights in the ECI were held constant over this period. Multiple studies suggest that the earnings of workers displaced in such sectoral changes declined as much as 25 percent in subsequent years (Jacobson, LaLonde, and Sullivan, 1993; Davis and von Wachter, 2017; Couch and Placzek, 2010), but these highly relevant changes in civilian earnings opportunities are likely to be attenuated by the fixed weight approach used in the construction of the ECI.

Inability to Measure Subgroup Earnings

A final limitation of the ECI relates to its flexibility in considering subgroups of workers rather than subsets of industries, occupations, and job types. The ECI publishes industry- and occupation-specific indices, but they do not account for compositional differences in the age and educational attainment of civilian workers, as highlighted above. Thus, the ECI may mask significant differences in the path of earnings for certain groups of workers. To demonstrate these differences, Figure 4.4 shows the path of earnings indices over a 37-year period generated using data from the CPS for all workers and for two subgroups defined by specific levels of educational attainment, those with a high school diploma or some college, and those with a master's degree or greater.[9] Like the ECI, these earnings indices track the change over time in earnings from some chosen base year—1982 in this case, which is set equal to 100—and follow the growth in relative earnings over time.[10] As can be seen, earnings growth among the full sample does not differ systematically across the time series from the earnings of workers with a master's degree or higher, even though these better educated workers make up only 9 percent of the sample. Their earnings disproportionately affect the average rate of earnings growth overall because, on average, they earn 200 percent of the earnings of workers with lower levels of educational attainment. In contrast, the growth in earnings over time for workers with a high school diploma, some college, or an associate's degree—who constitute 60 percent of the sample—grew at a significantly lower rate than the index using the full CPS sample, though all these workers are included in the full CPS sample index. Such differences in subgroup earnings growth suggest that compositional factors may be critically important in assessing the relevance of an index of earnings growth for the civilian opportunities of per-

---

administrative earnings records, suggesting that they might be more accurate than the individual, self-reported earnings in the CPS. However, we use a CPS respondent's report of weekly earnings from the ORG, rather than annual earnings from the CPS's Annual Social and Economic Supplement (ASEC), a measure that is less likely to suffer from recall and measurement error.

9   We show indices for other educational subgroups later in the Chapter in Figure 4.5.

10   Each index in this figure is a chained index, constructed similarly to the DECI (as described in more detail in the text below), with the only difference being that these indices use the sample weights in the CPS rather than custom weights reflecting the age and educational attainment composition of personnel in the military.

Figure 4.4
Comparison of Average Earnings of Full-Time Workers with Earnings of Two Educational Attainment Subgroups

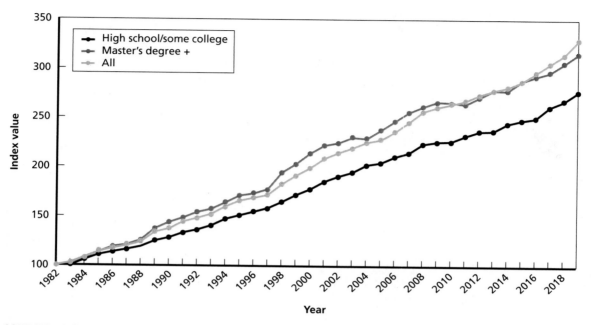

SOURCES: Civilian labor force educational attainment composition uses CPS ORG data (April–September, 1982–2019) from IPUMS.
NOTE: "Full-time" refers to workers who report usually working 35+ hours per week.

sonnel in the military—where the vast majority of service members hold either a high school diploma or have college credit short of a bachelor's degree—but the ECI is not designed to reflect such worker characteristics.

These aspects—the differing age, educational attainment, and gender composition of military personnel, the masking of business cycle effects on the rate of earnings growth associated with using job-specific data with fixed employment weights, and the inability to generate earnings measures for important subgroups of workers—motivate interest in an alternate earnings index.

An Alternative Measure: The Defense Employment Cost Index (DECI)

As discussed in Chapter One the DECI was first developed in the early 1990s at RAND as an alternative to the ECI for use in setting the annual military pay adjustment.[11] In this subsection, we first discuss the multiple features of the DECI that differentiate it from the ECI and make it a promising candidate for use in guiding the annual pay raise and monitoring other aspects of military pay. We then explain the construction of the DECI, including the data sources used and compare our approach to that used in the original Hosek 1992 study.

---

[11] Hosek 1992 introduced the DECI and Hosek, Peterson, and Heilbrunn (1994) extended the analysis of pay gaps between the DECI and ECI using additional years of data.

Advantages of the DECI

First, the DECI uses individual-level, worker-based survey earnings data from the CPS. The CPS sample includes around 130,000 individuals at any given time. These individuals are surveyed monthly, and approximately 80,000 respondents provide detailed information on their earnings in the week prior to being surveyed. The CPS uses an address-based sampling frame to contact individuals regardless of where they work and then follows them over time, regardless of whether or where they are employed. We provide more background on the CPS and the specific subsample of the survey we use in our discussion of the construction of the DECI below.

Second, the weighting of the DECI better represents active duty military personnel characteristics. The primary innovation of the DECI relates to its earnings weights. The weights in the DECI are derived from administrative data on current, active duty personnel in the four services.[12] For the basic DECI, we use eight age groupings, four educational attainment groupings, and gender to generate a set of 64 weights, one for each age-by-education-by-gender cell. As we will demonstrate below, this relatively basic division of people according to just three demographic characteristics generates a significantly different measure of the growth rate of earnings than the ECI and generally outperforms the ECI in predicting important measures of enlistment and continuation in the active duty military.

Additionally, the weights used for the DECI are updated annually, so that each year's earnings are weighted by the prior year's military composition. Updating these earnings weights annually (described in more detail below) means that the change in earnings measured by the DECI reflects not only changes in the actual earnings of civilian workers as they move from job to job, but also any changes in the age, educational attainment, and gender composition of active duty personnel that would affect the relevant path of earnings growth.

Third, the DECI is a framework that allows for the generation of subgroup DECIs that may be useful for various tasks. The DECI can be customized to index age-, education-, or gender-specific earnings for subgroups such as enlisted members or officers, specific services, or occupational specialties, and combinations of these groups. Such flexibility not only makes the DECI valuable in setting the annual pay raise but may also allow it to contribute to other aspects of military compensation, such as the differential use of enlistment or reenlistment bonuses and career pays across services and occupations.

---

[12] We generated weights for the DECI using active duty enlisted members and commissioned officers. We exclude warrant officers and Coast Guard personnel. We also exclude Navy commissioned officers because of a significant incidence of missing educational attainment data in the ADMF. These missing data were concentrated disproportionately among Navy officers with fewer YOS, especially in more recent years of data. Dropping these observations and including Navy officers could bias the overall DECI because these data would disproportionately represent more-senior Navy officers with higher civilian earnings opportunities. That said, in sensitivity analysis, we found that including Navy officers did not meaningfully affect the DECI results in either a qualitative or a notable quantitative fashion. Note that in contrast to the construction of the DECI, in the analysis in Chapter Three, we included Navy officers in the construction of the RMC weights. This is because the RMC analysis stratified the data by YOS, so we expect the bias to play less of a role unless missing education is non-random within a given YOS or YOS group. As a sensitivity analysis for Chapter Three, we computed the weighted average of RMC excluding Navy officers (as we do in the computation of the DECI) and found no qualitatively meaningful change in our results. We note also that, while we have found these issues to be relatively unimportant in our estimates, if they were of concern for DoD, the use of SOFS survey data in generating a DECI is also feasible, either instead of or in tandem with these administrative data.

Defining the DECI

The earnings data used in the DECI come from the Earnings Supplement, an additional set of questions asked of the Outgoing Rotation Group (ORG), a subsample of CPS respondents who are in either the 4th or 8th month of the period during which they are surveyed.[13] The measure we use is a respondent's self-report of weekly earnings at the job(s) they currently hold. The Earnings Supplement subsample includes wage and salary workers and excludes self-employed individuals, regardless of whether their businesses are incorporated. For each survey respondent, we additionally observe the maximum completed level of education when surveyed and their age, which we use to group respondents into eight grouped age cells. These age groupings are 17–21, 22–26, 27–31, 32–36, 37–41, 42–46, 47–51, and 52+. We also group respondents into one of four educational attainment cells. These are individuals with less than a high school diploma, a high school diploma or some college (including an associate's degree), a baccalaureate degree, and a master's degree or greater. In total, we have 32 age-by-education groups. We then average the earnings of CPS respondents within a given age-education-gender-year data cell using the provided CPS sample weights for the ORG subsample when generating within-cell average earnings.

To generate a timely measurement of earnings for the purpose of guiding the military pay raise decision, we use CPS ORG survey responses from the months of April through September. As discussed in Chapter One, the ECI from the third quarter of a given year are used to guide the military pay raise 15 months later. For example, the ECI for the third quarter of 2019 is used to guide the pay raise implemented in January 2021. In this example, for the DECI, the CPS ORG data for April through September of 2019 would be used to guide the 2021 military pay raise. Thus, the choice of the April–September subsample leads to about the same time-lag for the DECI as currently exists for the ECI. This sample restriction, along with a few others discussed in more detail below, results in a sample size of around 45,000 individuals per year, with a mean age-by-education-by-gender-by-year cell size of 1,376 respondents, a median of 1,303, and a standard deviation of 882 respondents.

The data used to construct the earnings weights are drawn from DMDC's ADMF. These administrative data contain information on service, enlisted or officer status, age, and educational attainment for all service members in the active duty force for the years 1982 to 2018.

The DECI is constructed as a chained Laspeyres index. This means that, rather than holding employment weights (as well as baseline earnings) fixed at some specified period, these elements are updated annually, so that the index compares the change in earnings between the current year and the prior year, rather than between the current year and a fixed baseline year. Thus, the DECI for 2019 compares the increase in civilian earnings from 2018 to 2019 for the age and educational attainment composition of active duty military personnel in 2018, but the

---

[13] The CPS survey uses a "panel" or longitudinal structure, meaning that individuals are kept in the sample and surveyed multiple times over a period of 18 months (specifically, for four initial months, then again for the same four months one year later). Those who are in their 4th and 8th months in the survey (the final months of each of these two periods) are referred to as the "Outgoing Rotation Group" (ORG) and these individuals provide answers to several additional questions including earnings from their current job(s). For more on the panel structure of the CPS and other details regarding the program, see Rivera Drew, Flood, and Warren (2014). Because some individuals appear in our annual ORG subsamples twice (due to their 4th and 8th months in the survey both falling between April and September, the months we use in the DECI), we avoid implicitly "upweighting" these individuals by randomly dropping one of these two observations from the data.

DECI for 2020 will compare the increase in civilian earnings from 2019 to 2020 for individuals matching the composition of active duty military personnel in 2019.

We formally illustrate the construction of the DECI for 2019 with 2018 as the baseline year. In this case, the ratio for the change in earnings from 2018 to 2019 used to calculate the DECI is identical to the ratio used in the ECI. In other words, for earnings, e, and earnings weights, w, the change in the DECI from 2018 to 2019 is defined as

$$\Delta DECI_{2018-2019} = \left[\frac{\sum_i w_{i,\,2018} e_{i,\,2019}}{\sum_i w_{i,2018} e_{i,2018}} \right].$$

But, unlike the ECI, this relationship holds for any other pair of years. Thus, $\Delta DECI_{2017-2018}$ substitutes this pair of years for 2018 and 2019. The DECI index is made up of the product of these year-to-year changes and a value of 100 for the (arbitrary) base year chosen. Thus, the DECI for the year 2019, with 2015 as the base year is defined as

$$DECI_{2019} = 100 \times \Delta DECI_{2015-2016} \times \Delta DECI_{2016-2017} \times \Delta DECI_{2017-2018} \times \Delta DECI_{2018-2019}.$$

It is important to note that for both the DECI and the ECI, the choice of baseline year (the year that is set equal to 100) is arbitrary. However, when comparing the path of indices over time, this choice can have a significant influence on what such a comparison suggests about both the sign and the magnitude of the difference between indices in any given year. Thus, the choice of baseline year is a subjective one that must be based on institutional knowledge or the context of a specific policy question. A useful candidate for the baseline year in comparing the DECI with the ECI, for example, would be a year in which policymakers considered the level of pay to be "correctly" set with respect to a policy goal (e.g., equal to the 70th percentile of civilian earnings). From this point, comparisons of differences between indices are meaningful in the context of the specific policy question "What is the appropriate path of annual pay adjustments given that pay was set correctly in year X?" We empirically explore the consequences of using different baseline years in the exposition that follows. That said, we use 1982 as a base year because, as argued in Hosek 1992, two large military pay increases occurred in 1981 and 1982 that reset military pay so it would be adequate relative to civilian pay to attract and retain the personnel the services required.

The Original DECI Study and Our Approach
Hosek 1992 introduced the DECI and conducted a series of novel analyses that explored the military/civilian earnings gaps predicted by the DECI and compared these predictions with those predicted by the ECI. Hosek 1992 also considered the differences in earnings gaps for a variety of subgroups of military personnel (e.g., by age, years of service, education), and showed a series of analyses to validate the DECI by comparing its path over time with the path of key measures of accession and retention. While much of our approach here follows Hosek 1992 closely, we made some changes in our specification of the DECI. Some of these choices were necessitated by differences in data access and scope of work, but some are based on considerations related to, for example, updated research or changes in data access in the intervening

decades. Below, we outline the most important ways in which our formulation of the DECI differs from the approach in Hosek 1992.

Inclusion of Gender in the Construction of the DECI

The issue of how to handle gender in constructing earnings comparisons, particularly with respect to guiding military pay, is not clear-cut. Several arguably relevant factors bearing on this decision can be used to argue for either including or omitting gender:

- the military pay schedule is gender neutral
- the level of male civilian earnings, after controlling for age and education, has historically been higher than female earnings and remains so today, while the rate of growth of female earnings has been higher than that of males in recent decades, narrowing this gap
- males make up around 85 percent of military personnel
- the share of women in the military has grown by more than 50 percent (from 9 percent to 16 percent) over the past four decades.

Hosek 1992 chose to use age, education, and occupation as the categories to include in the construction of the "full" DECI (the analogue to the ECI), omitting gender. They did, however, generate gender-specific subgroup DECIs in their analysis.

Our decision was to include gender in the construction of our main DECI, though we omit it in some subgroup DECI analyses in order to preserve a sufficient sample size in each data cell. Among the reasons we chose to include gender were

- to facilitate comparability with the RMC percentile analyses in this study that uses gender weighting (as well as other past military compensation work using the same approach)
- to allow the DECI to reflect the fact that many occupations exhibit high (but generally declining) levels of gender segregation (Cortes and Pan, 2017), meaning that weighting civilian earnings by the gender composition for the labor force partially serves as an implicit, and perhaps more realistic, control for occupation
- to allow changes over time in the share of women in the military to be reflected in the guidance for the appropriate annual basic pay raise.

We also, however, present results for a non-gender-weighted DECI in Appendix D and briefly discuss the differences in these indices. Further, we note that while our coding of the DECI in Appendix E uses gender, age, and education to define the weights, future iterations by DoD could change this code to reflect different decisions about the use of gender, age, or education as weights.

Omitting Occupational Groupings in the Construction of the DECI

A natural question to ask might be why we do not use occupation-specific earnings data to generate age-by-education-by-occupation cells for each year, so that our earnings measures could then be matched to military occupational specialty (MOS), perhaps increasing the fit between active duty members and their potential earnings in the civilian labor market. We chose to limit the DECI to age, educational attainment, and gender groups for two reasons, one statistical and one conceptual.

Statistically, adding further group types to the DECI subjects our estimated earnings data to the "curse of dimensionality," a concept in statistics that refers to the exponential growth of the total number of groups as additional criteria are added to a group structure, and to the way that this growth can quickly exhaust the ability of even very large datasets to provide sufficient numbers of observations to generate an accurate estimated outcome for each resulting subgroup (or to even have a single observation per group in many cases). Adding even very coarse occupational groupings would dramatically increase the overall number of groups in our analysis. Hosek 1992 used six occupational groups, which would increase our total number of groups from 64 to 384.[14] While, conceptually, these age-education-gender-occupation groups might more accurately describe the distribution of earnings in the labor force, in any given year we would be much more likely to encounter age-education-gender-occupation data cells with no individuals in them, which would necessitate the use of regression techniques to generate predicted values of these missing earnings. We viewed the accuracy of earnings estimates within each group as a more important goal than a multiplicity of groups that might better conceptually capture the full variance of earnings but would rely more heavily on prediction for the earnings measures themselves.[15]

Conceptually, the incorporation of occupation adds an element of speculation to the DECI which is otherwise absent. All individuals have an actual age, gender, and level of educational attainment that we measure with certainty, allowing a deterministic linkage between individuals in our data and a specific age-by-education-by-gender group. Adding occupation involves assuming one counterfactual occupational group for each active duty service member in our administrative data. The validity of such an exercise hinges critically on the empirical accuracy of the mapping of military occupation to civilian occupations (i.e., how likely is a service member in occupation j to later enter civilian occupation k). While specialized occupations such as aviator or numerous health care specialties may have clear analogs in the civilian labor market, there is a dearth of empirical evidence showing a strong connection between occupations such as Army infantry (Military Occupational Specialty 11B) and subsequent civilian occupation, and military occupations such as 11B make up a large portion of the overall force.[16]

---

[14] Hosek 1992 used only three educational groupings instead of our four groups, so the total number of groups in their analysis was 8 times 3 times 6, or 144, which is still around 2.5 times as many groups as we include in this analysis.

[15] For the small number of age-by-education-by-gender-by-year cells where we have missing earnings, i.e., cells with no individuals in the CPS subsample from which to draw an earnings self-report (which is limited to the subcategory "Ages 17–21 with a Master's Degree or Higher," a data cell that is empty for 19 of 37 years among males and 11 years among females) we estimate the following regression model and use the predicted values from this regression to impute these missing earnings:

$$\log(earnweek_{ist}) = \alpha_1 age_{it} + \alpha_2 educ_{it} + \alpha_3 female_i + \alpha_4 year_t + \beta_1(age_{it} \times educ_{it}) + \beta_2(age_{it} \times female_i) + \beta_3(age_{it} \times year_t) + \beta_4(educ_{it} \times female_i) + \beta_5(educ_{it} \times year_t) + \beta_6(female_i \times year_t) + \varepsilon_{it}.$$

In contrast with the CPS data, for each year of ADMF data on military personnel, there are between 1 and 114 individuals with this combination of age and educational attainment in all but one year in our data. However, given the weighting implied by the extremely small size of these cells, the role of these few imputed earnings in our overall results is highly trivial. (Though, a bit ironically, the generation of these imputed values is the most computationally intensive aspect of our analysis.)

[16] Assessing the connection between military skills and civilian careers is an area of active research (see, e.g., Wenger et al. [2017] and Hardison et al. [2017] for research comparing current and alternate approaches to mapping both technical and nontechnical skills acquired in the military to civilian careers).

Finally, from a practical standpoint, occupational linkages were not critical to the general findings of Hosek 1992. The authors tested the sensitivity of omitting occupation, estimating a DECI with only eight age groups and three educational attainment groups, and found that excluding occupation made little difference in the estimated pay gap (the largest difference in the size of the pay gap in any single year was around 11 percent).

We note that while we construct a DECI in a way that circumvents potentially problematic occupational mappings, there may be cases when well-defined occupational linkages between military occupation and civilian occupational opportunities would be of interest, such as for aviators or health care professionals. In such cases, nothing in our approach precludes the generation of a DECI for a subsample of military personnel using civilians within a certain set of occupations if the relevant civilian subsample is sufficiently large.

Differences in Educational Attainment Categories

We expanded the educational attainment categories used in Hosek 1992 by adding a fourth educational category, master's degree or greater. The original study classified active duty service members into three education categories: less than a high school diploma, a high school diploma only, or college. As shown in panel B of Figure 4.2, the share of individuals with a master's degree or higher nearly doubled between 1982 and 2018. Furthermore, as we show below, these individuals are clearly on a different path of earnings growth over time.

We were unable to ascertain whether individuals with an associate's degree or those who completed some amount of college coursework were placed into the high school diploma only category or the college category in Hosek 1992. For our analysis, we collapsed individuals with a high school diploma, some college, or an associate's degree into one category.[17] Figure 4.5 shows earnings sub-DECI indices by educational attainment. Aside from a modest divergence in the early 1990s, the paths of the sub-indices for those with only a high school diploma and those with some college or an associate's degree are virtually identical over nearly 40 years. The figure also shows that each of the other groups we use (less than high school diploma, bachelor's degree, and master's degree or greater) follow a distinct, non-overlapping path over the entire time series, suggesting that these four categories of educational attainment capture the most relevant associations between educational attainment and earnings growth.[18]

Choosing the CPS Subsample

Hosek 1992 utilized CPS data from the month of March. Each year, a subsample of March CPS respondents is asked to complete a considerably longer survey known as the Annual Social

---

[17] As we discussed in Chapter Three, we observed discrepancies in the proportion of service members with educational attainment levels of high school diploma, some college, and associate's degree in the ADMF versus the SOFS. When we aggregate these categories together though, we find that the results from the ADMF are highly comparable with the SOFS data on educational attainment for the years of overlap, giving us confidence in the accuracy of this aggregated category for use in constructing the DECI. In Appendix D, we present an alternate DECI generated from SOFS tabulations for the years 2002–2018 using six educational attainment levels (less than high school, high school graduate, some college, associate's degree, bachelor's degree, and master's degree or greater). This index doesn't differ meaningfully from the DECI generated using ADMF data and four educational attainment categories.

[18] We are also unable to ascertain the specific definitions of less than high school versus high school diploma holders in Hosek 1992. Our definition of "less than high school" includes attending high school as a junior or less; attending high school as a senior; secondary school credential near completion; test-based equivalency diploma; occupational program certificate; correspondence school diploma; high school certificate of attendance; home study diploma; and adult education diploma. Our definition of "high school diploma" includes traditional high school graduates and GED, and Army National Guard Challenge Program degree holders.

Figure 4.5
Comparison of CPS Earnings Sub-Indices by Education Group

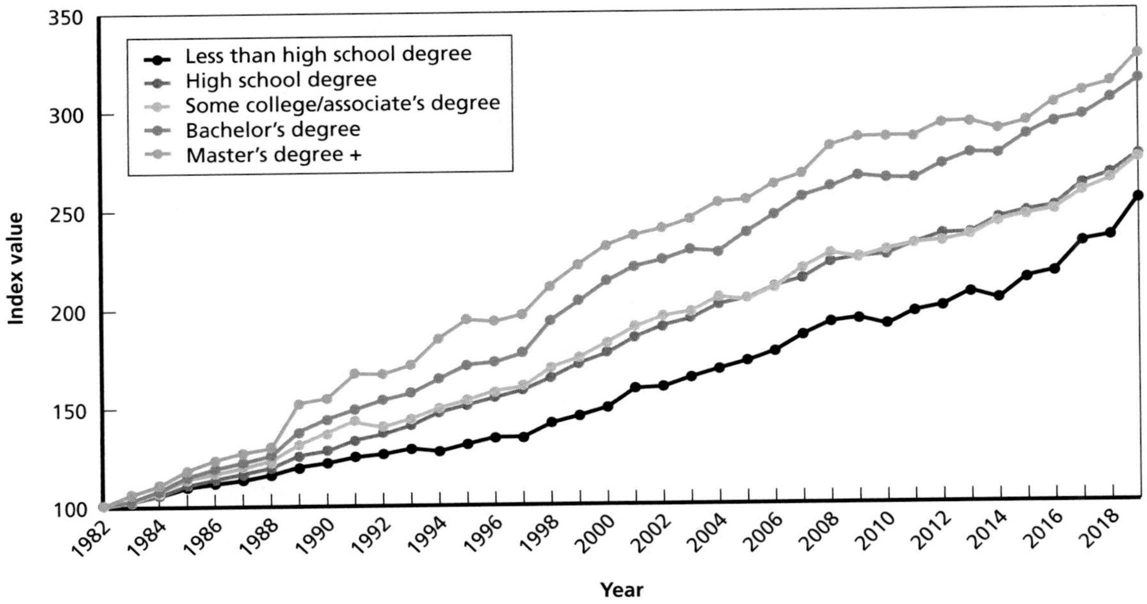

SOURCE: Civilian labor force educational attainment composition uses CPS ORG data (April–September, 1982–2019) from IPUMS.

and Economic Supplement (ASEC), also commonly referred to simply as the "March Supplement." In this additional questionnaire, respondents are asked to estimate their annual earnings over the prior 12-month period. Thus, the DECI estimated by Hosek 1992 was based on data from the 12 months prior to the March survey, implying that if the DECI computed in this way was used as a guide for, say the January 2021 military pay raise, it would be based on the March 2019 ASEC, covering the period April 2018–March 2019, a substantial time lag. Recognizing the potential importance of this time lag, Hosek 1992 stated in a footnote (p. 13) that they "used the March CPS income data because a suitable data file was available at RAND when the project began." They went on to suggest that using weekly earnings data from the CPS ORG questionnaire could improve on the timeliness of the DECI.[19]

We follow this suggestion and use self-reported weekly earnings from respondents in the ORG subsample between the months of April and September. We use this six-month subsample of ORG respondents for multiple reasons. First, these months are safely removed from the traditional cycle of seasonal employment that lasts from roughly November through January (see Alhassan [2019] for more on the nature of this annual employment cyclicality). Second, the respondents across this period represent a subsample large enough to generate stable earnings estimates even when considering subgroups (the average sample size of our six-month ORG subsample is approximately 28 percent larger than the size of the ASEC subsample on

---

[19] Data availability was a nontrivial issue when the Hosek 1992 study was conducted. In the ensuing decades, both computing power and data availability have improved vastly, making the type of analysis we conduct—explicitly comparing results using two CPS subsamples—relatively trivial. Had the authors of the original study been able to take advantage of these improvements, we are certain they would have pursued this same course.

average). The third reason is the timeliness issue mentioned in Hosek 1992.[20] Finally, in the series of validation exercises we conducted discussed later in this chapter, we found that the DECI using the ORG data consistently performed at least modestly better than the DECI using ASEC data (results using the ASEC are not shown later in the chapter).

Comparing Relative Pay Growth Using the ECI and the DECI

To assess the trends in earnings growth estimated by the ECI and the DECI, we use a series of graphical results to compare the path of each index over time to the path of an index of the growth in basic pay and discuss the extent to which the growth of military basic pay has diverged with the growth of civilian earnings in each comparison.[21] The basic pay index (BPI) is the measure of the growth in basic pay and is generated by starting with a value of 100 in the chosen baseline year and then multiplying this value by the percent change in basic pay in each subsequent year.[22] In addition to highlighting the differences in the ECI versus the DECI, we also stress that the guidance provided by these comparisons depends critically on the choice of baseline year. The importance of this aspect of the analysis is difficult to overstate. Indices, by their nature, can only provide guidance on rates of change, and have nothing to say regarding the comparability of earnings in levels. The "correct" baseline year is a policy choice that must reflect the views of decisionmakers about when the relative level of military pay to civilian earnings was set appropriately.

We focus our graphical analyses around two baseline years that tell significantly different stories about the relative paths of civilian and military earnings. The first is 1982, as was used in Hosek 1992.[23] This was the year after two "catch-up" pay increases—implemented in October 1980 and October 1981—that were intended to restore basic pay to a level considered appropriate for the military to be competitive with the civilian labor market. The second year we use is 2010, which was the final year of an 11-year period of continuous increases in basic pay equal to the ECI plus one half percent. These increases were mandated by statute from 2000 to 2006 in order to correct a perceived pay gap associated with the growth in civilian earnings over the period of the dot-com boom in the late 1990s. They were then extended for another five years to support retention and recruiting through the heaviest years of Operation Iraqi Freedom/Operation Enduring Freedom. The end of this unique period of pay increases and heavy deployments overseas makes this a plausible year from which to benchmark more near-term pay growth comparisons.[24]

---

[20] Further discussion of these differences, including a figure illustrating the monthly differences in surveying of respondents and their period of earnings measurement are in Appendix D.

[21] In Appendix E, we provide tables with the underlying data points for these and other figures in this report.

[22] Thus, for a 3 percent increase in basic pay in the year after the chosen baseline year, the BPI would be 100.00*1.03 = 103.00. For a 5 percent increase in the following year, the BPI would increase to 103.00*1.05 = 108.15, and so on.

[23] More specifically, we use the 12-month percent change in the ECI from end of September 1981 to September 1982, representing the earnings change from the end of FY 1981 to the end of FY 1982 to begin our series. Our ECI values are from a continuous series from 1976 to 2019 measuring end-of-September 12-month percentage changes that can be found in Table 9 of Bureau of Labor Statistics, National Compensation Survey (2020).

[24] However, it is worth noting that, because of the sustained increases in basic pay over the 2000s, RMC was at or often well above the 80th percentile by 2009 for most personnel across a variety of comparisons by service and by education level, as discussed in Chapter Three. Thus, this choice of benchmark satisfies a potentially important criterion concerning

Figure 4.6 plots the time series of the BPI, the ECI, and the DECI from the baseline year of 1982. Perhaps the most striking feature of this comparison is that beginning immediately after the baseline year, the ECI jumped to a higher value than the BPI and this implied earnings gap grew consistently until 2000, when the decade of pay increases discussed above fully closed the gap by 2010. The effect of the Great Recession can be detected in a modest downward inflection in the ECI between 2009 and 2010, at which point the value of the BPI exceeded the ECI for the first time in more than 25 years and remained above it for around five subsequent years. Since 2016, the ECI again exceeded the BPI by a small, but growing, amount.

The earnings gap revealed by comparing the DECI to the BPI in Figure 4.6 bears almost no resemblance, even in a qualitative sense, to that suggested by comparing the ECI to the BPI. For two decades after the baseline year of 1982, there was no systematic evidence of a pay gap using the DECI, in contrast to the ECI-based comparison. From 2000 to 2010, the above-ECI increases in basic pay caused the BPI to diverge dramatically from the path of the DECI. By 2008, at the onset of the Great Recession, the growth in basic pay had opened up a large gap between the BPI and the DECI, reflecting slower pay growth for civilians relative to military personnel. During the period of the Great Recession and its aftermath, the gap between the BPI and DECI grew even larger, reflecting stagnation in civilian earnings growth likely driven

Figure 4.6
Comparing ECI- and DECI-Based Civilian Earnings Growth over Time Using 1982 as the Baseline Year

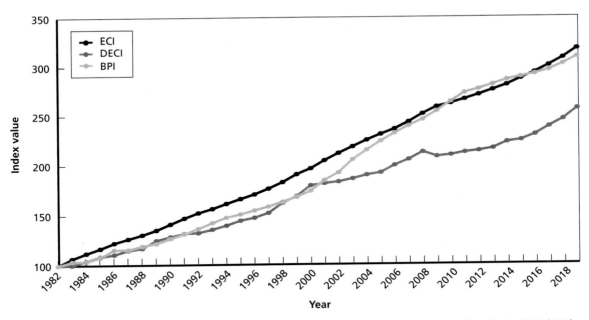

SOURCES: CPS ORG data (April–September, 1982–2019) from IPUMS, ECI data from BLS, BPI data from OUSD(P&R) (2018).
NOTES: DECI cells comprise eight age groups, four education groups, and gender. Military weights are generated from ADMF data aggregating officers and enlisted from four services but omitting Navy officers because of poor coding of education.

policy stability but does not meet the "70th percentile" criterion for the desired absolute level of military pay relative to civilian earnings.

by a combination of continuously employed workers not receiving pay increases, and millions of other workers having to find new jobs as their existing jobs were eliminated, which likely resulted in many workers receiving a nominal cut in earnings. At its largest, in 2011, the value of the positive gap between the BPI and the DECI was 24 percent of the overall index value.

To better show the implications of the ECI or the DECI for gauging the growth in basic pay, we follow Hosek 1992 and plot a standardized measure of the gap (or surplus) in the difference between BPI and either the ECI or the DECI. If the ECI is the civilian pay index of interest this earnings gap measure is defined as (BPI – ECI)/BPI. This normalized ratio measures the distance between the BPI and the comparison index (ECI or DECI) as a proportion of the level of the BPI (i.e., percent divided by 100). Figure 4.7 displays these differences as earnings gaps or surpluses between the BPI and the relevant civilian index over time, normalized by the value of the BPI. The ECI-based differences (gray bars) are all negative (earnings gaps) across the 1980s into the early 2000s and are often as large as 10 to 12 percent before approaching zero or reflecting an earnings surplus during the last decade. The DECI-based earnings differences (blue bars) are centered around zero and are generally smaller than 5 percent in absolute magnitude prior to the early 2000s. The relatively small gap in BPI versus DECI suggests that basic pay grew at a roughly similar rate as the pay of civilians with similar characteristics as military personnel. But after the early 2000s, BPI far exceeded the DECI, as the series of basic pay increases during the 2000s led to a sustained earnings surplus relative to civilian earnings that exceeded 20 percent for multiple years during the early 2010s.

Under the assumption that the level of military pay was set correctly relative to civilian earnings in 1982, the policy guidance implied by these two comparisons is starkly different.

Figure 4.7
Comparing DECI- and ECI-Based Earnings Differences Using 1982 as the Baseline Year

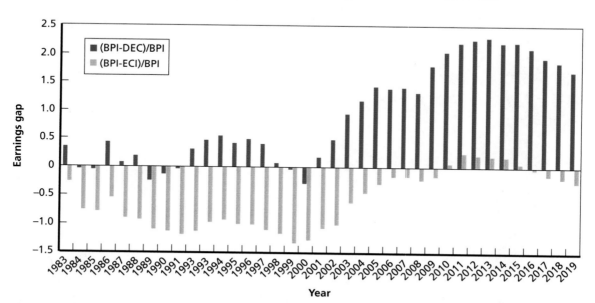

SOURCES: CPS ORG data (April–September, 1982–2019) from IPUMS, ECI data from BLS, BPI data from OUSD(P&R) (2018).
NOTES: DECI cells stratified by eight age groups, four education groups, and gender. Military weights are generated from ADMF data by aggregating officers and enlisted, omitting Navy officers because of poor coding of education. The formulas for earnings gaps using each earnings index are given in the legend above and they are measured as a proportion (i.e., percent/100), so that 0.2 equals 20 percent.

The ECI suggests that basic pay growth consistently lagged the growth in comparable civilian earnings for more than 20 years until the early 2000s, while the DECI suggests that there was no systematic gap in the growth of military and comparable civilian pay through this period. After that time, the rate of basic pay growth significantly exceeded the rate of growth of relevant civilian earnings when measured by the DECI but not by the ECI. But how do these comparisons change using another year, 2010, as our baseline year?

Figure 4.8 and 4.9 present results for the time series of these indices and the estimated earnings gaps or surpluses, but with all three indices equalized to 100 in 2010. The immediate divergence between the BPI and the ECI reflects the final year of above-ECI pay growth in basic pay. From 2011 to 2014, the basic pay increase matched the ECI, followed by three years of below-ECI increases in basic pay, which led to the reemergence of an earnings gap in 2016. Since 2017, the basic pay raise again tracked the ECI, holding the estimated earnings gap proportionally constant.

Comparing the DECI with the BPI over this time indicates that, from 2010 to 2013, basic pay grew faster than comparable civilian earnings. After that time—which coincided with the end of a five-year decline in the labor force participation rate in the United States—civilian earnings measured by the DECI grew at a faster rate than basic pay. This resulted in an earnings gap from 2017 to the present that was, by 2019, approximately equal in size to the earnings gap implied by the ECI. Notably, however, projecting the trajectory of earnings growth since 2015 forward suggests that a higher growth rate of basic pay would be required to close this gap from the perspective of the DECI than the ECI. To further demonstrate the sensitivity of the guidance provided by both of these indices to the chosen baseline year, Table D.1 in Appendix

Figure 4.8
Comparing Earnings Growth over Time Using the ECI and the DECI Using 2010 as the Baseline Year

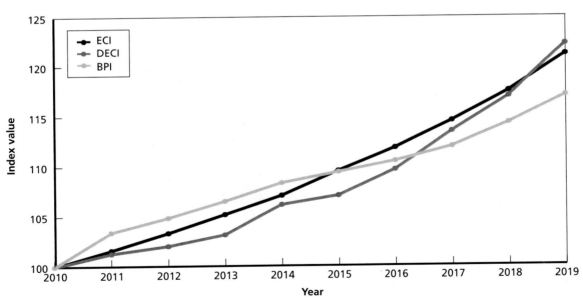

SOURCES: CPS ORG data (April to September, 1982–2019) from IPUMS, ECI data from BLS, BPI data from OUSD(P&R) (2018).
NOTES: DECI cells comprise eight age groups, four education groups, and gender. Military weights are generated from ADMF data aggregating officers and enlisted from four services but omitting Navy officers because of poor coding of education.

Figure 4.9
Comparing DECI- and ECI-Based Earnings Differences Using 2010 as the Baseline Year

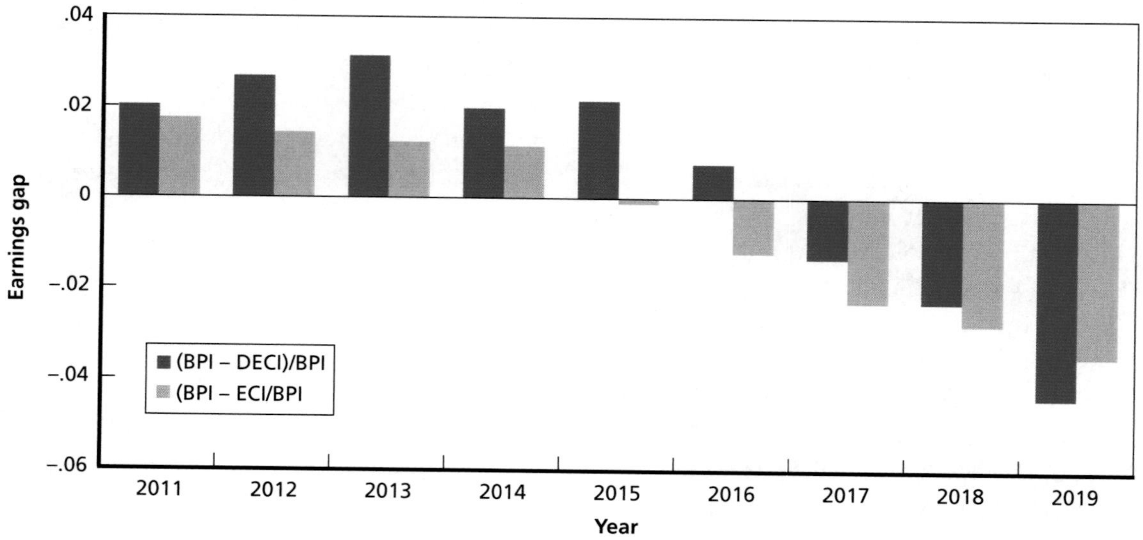

SOURCES: CPS ORG data (April–September, 1982–2019) from IPUMS, ECI data from the BLS, BPI data from OUSD(P&R) (2018).
NOTES: DECI cells stratified by eight age groups, four education groups, and gender. Military weights are generated from ADMF data by aggregating officers and enlisted, omitting Navy officers because of poor coding of education. The formulas for earnings gaps using each earnings index are given in the legend above and they are measured as a proportion (i.e., percent/100), so that 0.2 equals 20 percent.

D calculates cumulative percent changes in the BPI, DECI, and ECI using four different baseline years, 1982, 1990, 2000, and 2010.

It is worth reiterating that all of this discussion presupposes that each baseline year considered is "correct" in terms of absolute levels of pay and that subsequent differences are interpretable as either earnings gaps or surpluses. But note that this cannot be the case for both of the years we consider above as, for the DECI, the relative incidence of gaps or surpluses across this period are different for the two series. Interestingly, the 11-year period of above-ECI increases in basic pay from 2000 through 2010 had the result of driving the ECI-based earnings gap measured using 1982 as the baseline year back to approximately zero by the year 2010. This led the ECI-based series from 2010 to 2019 to be approximately identical regardless of which baseline year was chosen. In contrast, the DECI comparisons using 1982 as the baseline year suggested that, by 2010, service members received a large surplus relative to their civilian earnings opportunities, but this earnings surplus was set to zero in the second analysis using 2010 as the baseline year. From this baseline, we estimate much smaller earnings surpluses that turn into an earnings gap over the last three years of the time series. This qualitative difference in guidance implies that at least one of these combinations of indices and baseline years is providing more useful guidance than others. From the perspective of the "70th percentile" policy, evidence of a sustained earnings surplus from the early 2000s onward when measuring earnings growth using the DECI is consistent with both the findings of this study and of prior work (Hosek, Asch, and Mattock, 2012) suggesting that current RMC is closer to the 90th percentile of civilian earnings for many service members.

Using the DECI to Analyze Earnings Growth Among Subgroups

A key strength of the DECI is the ability to generate DECIs for specific subgroups of military personnel. This type of analysis can provide useful estimates of earnings growth for subgroups that may, for example, better represent comparable pay for officers versus enlisted personnel, specific pay grades or groups of pay grades, specific military occupations or groups of occupations, college educated versus high school graduates, or older versus younger service members. The information gleaned from such subgroup analyses may help to inform decisions on, for instance, the optimal path of earnings growth within the pay table or the use of SRBs, as noted earlier. We present examples for subgroups defined by age and educational attainment, as well as enlisted personnel and officers in Appendix D.

What Factors Drive the Differences Between the ECI and the DECI?

As shown above, there are large and qualitatively important differences in civilian earnings growth trends according to whether the ECI or the DECI is the measure used to estimate them. Earlier in the chapter, we highlighted two key conceptual differences that could contribute to the observed lack of agreement between these two indices: (1) the use of the CPS, an employee-based survey, versus the ECI, an employer-based survey, and (2) the weighting of civilian earnings by the age and educational attainment composition of individuals in the military. We now present evidence on the relative roles of each of these factors.

Differences in the Survey Data

Figure 4.10 plots three earnings indices over time: the ECI, the DECI, and a third CPS index that is calculated equivalently to the construction of the DECI but using the sample weights provided in the CPS data representing the U.S. population overall instead of the military-specific weights used in the DECI. Comparing the ECI with the CPS index over time, it is apparent that they are qualitatively similar. The two series trend strongly together over the 37-year sample period and end up with a difference of only around 3 percent by 2019. However, some of the differences in a given time period are not trivial. For example, earlier in the chapter we discussed the "pay paradox" implied by the ECI throughout the 1980s and into the early 1990s, the period analyzed in Hosek 1992. In Figure 4.10, it is apparent that the CPS index is notably lower throughout this period, even without the reweighting to reflect the composition of personnel in the military. Figure 4.11 presents the military/civilian earnings differences implied by the ECI, as shown in earlier figures. But instead of comparing these results with the DECI, they are compared with the earnings differences that result from using this simple CPS index. Across this nine-year period, the CPS earnings gap is around 60 percent of the magnitude of the gap implied by the ECI. This suggests that the nature of the underlying data and specifically the use of employee earnings data rather than employer cost data has important implications for the estimated path of civilian earnings growth.

Differences in Weighting

Because of substantial and persistent differences between military service members and the civilian labor force in both age and educational attainment, as shown earlier, it is not surprising that weighting the civilian labor force according to these characteristics contributes importantly to the large differences between the DECI and the ECI over time. The effect

Figure 4.10
Comparison of the ECI, the DECI, and the CPS Earnings Index

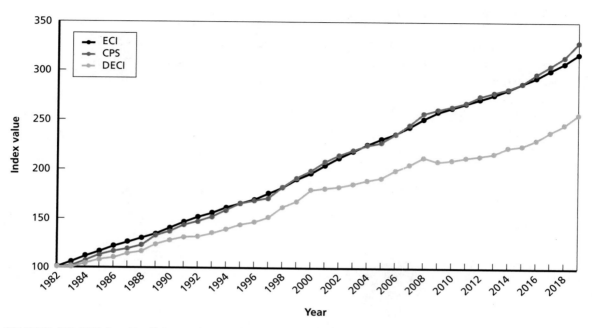

SOURCES: CPS ORG data (April–September, 1982–2019) from IPUMS, ECI data from the BLS.
NOTES: DECI cells stratified by eight age groups, four education groups, and gender. Military weights are generated from ADMF data by aggregating officers and enlisted, omitting Navy officers because of poor coding of education.

Figure 4.11
Comparing ECI-Based and CPS-Based Earnings Differences

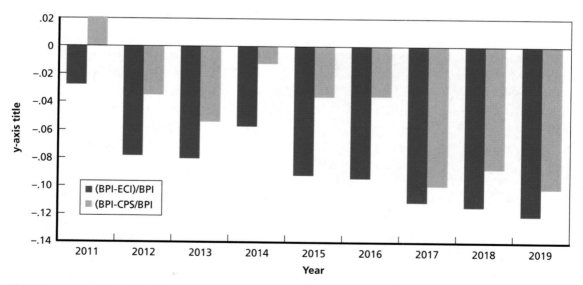

SOURCES: CPS ORG data (April–September, 1982–2019) from IPUMS, ECI data from the BLS.
NOTES: DECI cells stratified by eight age groups, four education groups, and gender. Military weights are generated from ADMF data by aggregating officers and enlisted, omitting Navy officers because of poor coding of education. BPI data from OUSD(P&R) (2018). The formulas for earnings gaps using each earnings index are given in the legend above and they are measured as a proportion (i.e., percent/100), so that 0.2 equals 20 percent.

of reweighting on the CPS data is visually apparent in Figure 4.10, in which the weights are entirely responsible for the different paths of the CPS index and the DECI. Figure 4.12 shows CPS sub-indices for the four aggregated age groups that we used earlier in Figure 4.1. Recall that nearly 90 percent of military service members are less than 38 years of age and around 60 percent are less than 27. The lower path of earnings growth for the two youngest age groups (ages 17–21 and 22–26) in these age-specific CPS indices makes clear that placing significant weight on these age subgroups, as is the case with the DECI, will lead to lower earnings growth. In fact, the overall path of the DECI closely approximates the paths of these two age subgroups, including the notable dip in relative rate of earnings growth during the years of the Great Recession and its aftermath.

Earnings differences by educational attainment also contribute significantly to differences between the DECI and the ECI. Figure 4.5 showed substantial divergence in earnings paths across educational categories over time. The gap in earnings growth by education was largest during the Great Recession and has narrowed somewhat in the post-recession years as the labor market participation of less educated individuals declined, thereby reducing the number of available workers and tightening the market for lower-skilled occupations. Figure 4.13 presents the ECI, the DECI, and two CPS subgroup indices for those with less than a high school diploma, and those with a high school diploma, some college, or an associate's degree. Together, these two groups make up approximately 80 percent of active duty personnel. Not surprisingly, the path of the DECI across nearly 40 years of data lies between the curves for these two educational attainment categories, suggesting that the DECI can also be well approximated as a weighted average of the earnings growth of workers in the civilian labor market with an associate's degree or less.

Figure 4.12
Comparison of CPS Earnings Sub-Indices by Age Group

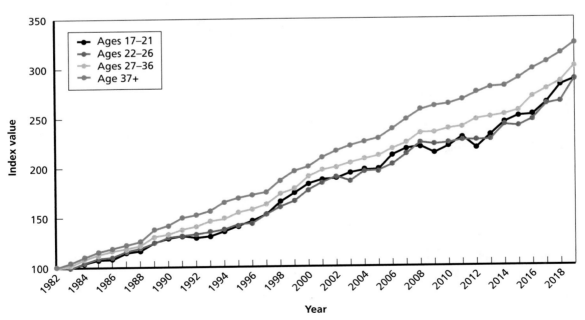

SOURCE: Civilian labor force age composition uses CPS ORG data (April–September, 1982–2019) from IPUMS.

Figure 4.13
Comparison of ECI, DECI, and CPS Earnings Sub-Indices by Education Group

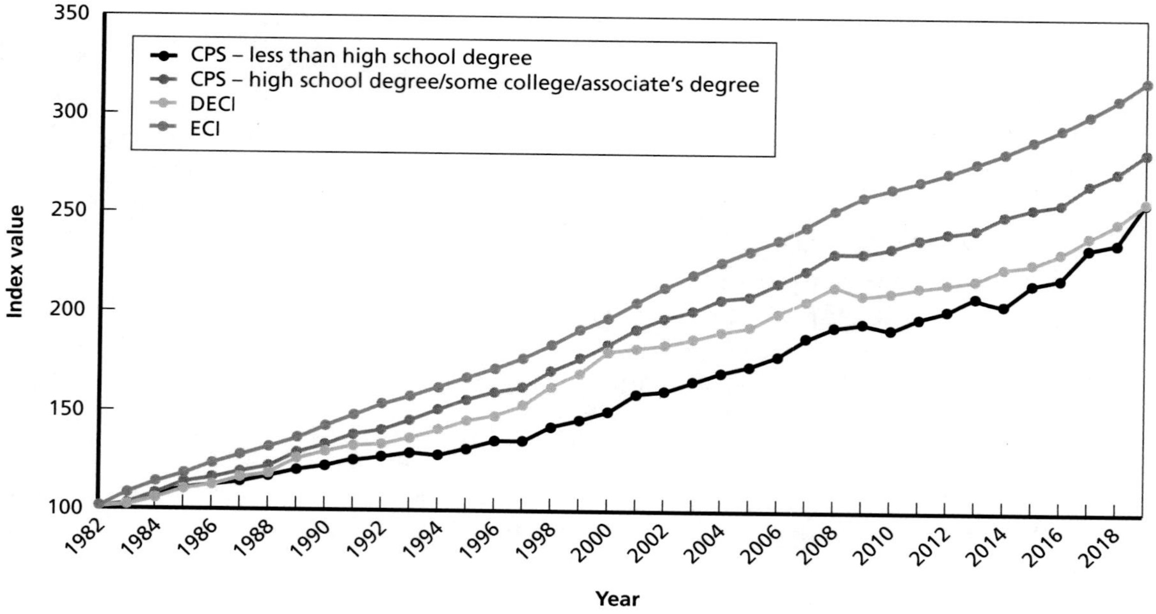

SOURCES: Civilian labor force educational attainment composition uses CPS ORG data (April–September, 1982–2019) from IPUMS. ECI data from the BLS.
NOTES: DECI cells stratified by eight age groups, four education groups, and gender. Military weights are generated from ADMF data by aggregating officers and enlisted, omitting Navy officers because of poor coding of education.

Finally, the role of gender weighting plays a role in the divergence of the DECI and a gender-neutral measure such as the ECI. Figure 4.14 shows CPS gender subgroup indices. Using 1982 as a base year, the rate of earnings growth has been consistently greater for women relative to men such that, by 2019, the index value for women is around 25 percent higher than the value for men. This is in contrast to the *level* differences in earnings, which are higher for men than women on average. The higher rate of earnings growth is what has driven a reduction in this level difference in earnings by gender over the past 40 or so years in the United States. Because the full-time labor force is much more close to gender parity than the military (around 44 percent of the full-time labor force is female versus about 15 percent of the military), weighting the DECI by gender has the effect of upweighting the rate of earnings growth among men, making the rate of earnings growth measured by the DECI lower than a gender-neutral index like the ECI.

How Strongly Is DECI Associated with Military Manpower Outcomes?

The factors that drive the differences between the ECI and the DECI give face validity to the DECI as a measure of the path of civilian earnings growth for those in the military. But the relative value of the DECI and the ECI for guiding the annual pay adjustment is ultimately an empirical question regarding how strongly the DECI versus the ECI correlate with key military manpower outcomes over time. We examine this question in this subsection by visually

Figure 4.14
Comparison of CPS Earnings Sub-Indices by Gender

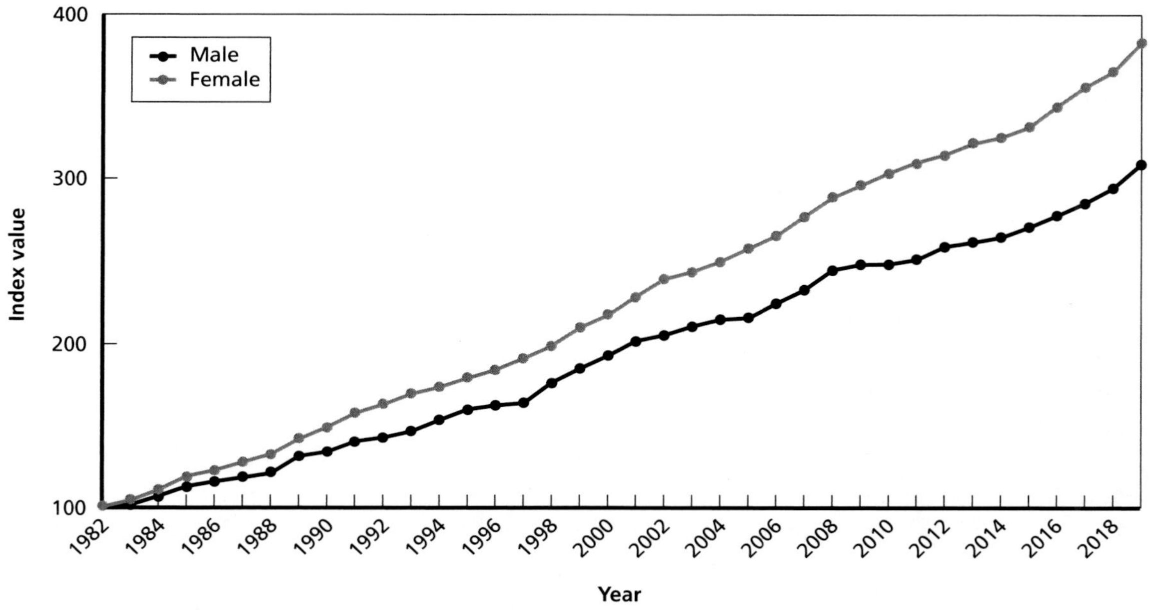

SOURCE: Civilian labor force age composition uses CPS ORG data (April–September, 1982--2019) from IPUMS.

and statistically comparing associations between the path of a ratio of the BPI and either the ECI or the DECI over time with the path of the percentage of high-quality accessions, enlisted continuation at 4 YOS, and officer continuation at 8 YOS.[25] We find that the DECI performs substantially better than the ECI in tracking high-quality accessions, while both indices track continuation less accurately overall and about equally well.[26]

We reiterate two important points made in Chapter Three. First, factors other than relative military growth affected these outcomes. Second, the analysis of the association between manpower outcomes and the ratio of BPI to ECI versus the ratio of BPI to DECI is not intended to imply that relative military pay is the only or even the most important policy tool available to the services for influencing outcomes. As noted earlier, other policies, such as bonuses, affect these outcomes and have been found to be more cost-effective policies than pay as a means of influencing these outcomes. But we focus here on BPI relative to the different indices of civilian pay growth because of our interest in providing additional information about the validity of the ECI versus the DECI as a guide for setting the annual increase in basic pay. Also, as noted before, manpower outcomes are affected by other factors, including service

---

[25] The definitions of high-quality accessions and continuation are provided in Chapter Two.

[26] We note that continuation at YOS 4 is not an ideal measure because individuals under a service obligation—such as those with a five or six-year service obligation are not free to make a retention decision. That is, the continuation rate reflects both the voluntary retention decisions for those not under a service obligation and the attrition decision for those under a service obligation. At least for enlisted personnel, a better metric would be the reenlistment rate among those eligible to make a reenlistment decision. The DMDC data we used did not have adequate information on those completing a service obligation and eligible to make a retention decision, so we used continuation data instead, given the timeline and scope of work for this project.

policies to influence these outcomes. Because we do not control for these factors or present a causal analysis, the associations we present should be viewed as descriptive only.

We use two exercises to assess the strength of associations between civilian earnings growth, as estimated by the DECI and the ECI, and these outcomes. The first is a simple graphical exercise that superimposes two curves across the nearly 40 years of our sample period to visually compare their association. To use the example of accessions, the first curve measures the percentage of all accessions that are high-quality (on the left-hand-side y-axis). The second is the ratio of the BPI to either the ECI or the DECI (on the right-hand-side y-axis). This ratio is 1 when the indices are equal, falls below 1 when the civilian earnings index has grown faster than the BPI, and grows larger than 1 when the BPI has grown faster than the civilian earnings index. Thus, the position of this ratio relative to a value of 1 indicates whether relative earnings growth from year to year has favored military service or civilian employment, and its magnitude indicates the strength of this relationship.[27] The second exercise is to estimate a bivariate ordinary least squares regression model that quantifies the statistical association between these measures. While we report both the slope coefficients for each regression, we focus on the R-squared value, which measures the proportion of the total variation in the military outcome of interest that can be explained by variation in the earnings ratio.[28] For each exercise, we use a subgroup DECI reflecting weighting that is more plausibly related to the outcome measure of interest.[29]

Figure 4.15 shows the trends in the subgroup DECI of enlisted personnel and the percentage of accessions in all four services that are high-quality.[30] From the mid-1990s on, both indices track high-quality accessions broadly in terms of whether both measures are trending upward or downward; however, from the early 1980s to the early 1990s, the ECI is strongly negatively correlated with percentage high-quality accessions (reflecting the "pay paradox" discussed earlier). Over this period, the percentage of high-quality accessions grew constantly from around 45 percent to 75 percent, while the BPI/ECI ratio fell from 1.00 to around 0.88. In contrast, over this same period, the BPI/DECI ratio grew by around 5 percent. In Panel A of Table 4.1, we present regression results for which the dependent variable is percentage high-quality and the explanatory variables in the two regression models are, respectively, the BPI/ECI and BPI/DECI ratio. Comparing the R-squared statistics from each of these regressions, we see that the BPI/DECI ratio explains nearly half of the total variation in the percentage of

---

[27] We use 1982 as a baseline year for these comparisons, but there are no meaningful quantitative differences in the results if we use 2010 as the baseline year since we are, in essence, comparing the overall shape of this curve rather than its value in any given year. To use a neutral approach to determine how these series are graphically overlaid on one another, we plot each series using a vertical range that spans plus or minus two standard deviations from the mean value of the series.

[28] We estimate a bivariate regression model of the type $military\_outcome = \beta_0 + \beta_1 earnings\_ratio + \epsilon$. In such a model, R-squared explains how much of the overall variation in the military outcome can be explained by the earnings ratio when we optimally set two parameters, a single overall level offset between the series (the constant term, β_0) and a single slope coefficient, β_1—where *optimally* means finding the parameters that minimize the sum of the squared differences between each series of data points.

[29] For the subgroup DECIs we use in this exercise (an enlisted DECI and an officer DECI), we do not weight by gender because of the cell size issues discussed earlier in the chapter.

[30] We also conducted this exercise using a subgroup DECI of only younger service members (26 and below) with educational attainment below a bachelor's degree, and the results were qualitatively similar to those presented here, due to the fact that younger service members represent the significant majority of enlisted personnel.

Figure 4.15
Comparison of Enlisted DECI and ECI with the High-Quality Accession Rate

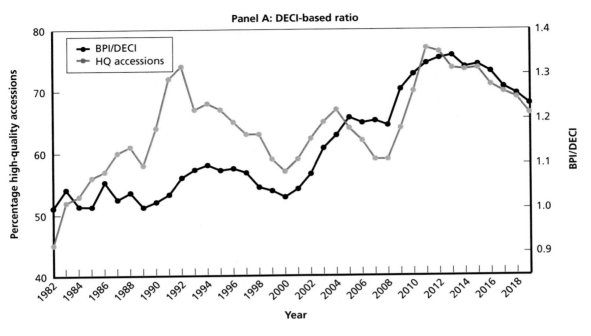

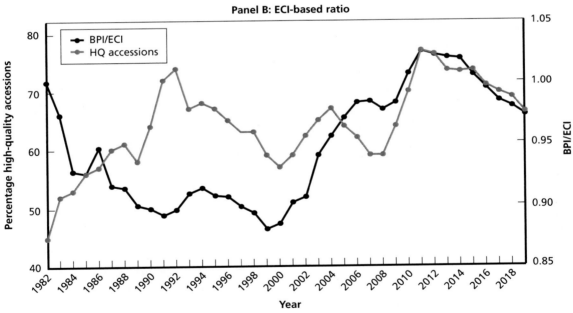

SOURCES: CPS ORG data (April–September 1982–2019) from IPUMS. ECI data from the BLS. Accession data from OUSD(P&R).
NOTES: DECI cells stratified by eight age groups and four education groups. Military weights are generated from ADMF data on enlisted personnel.

high-quality accessions, while the BPI/ECI ratio explains 11 percent. It is notable that the estimated coefficients are nearly identical, but the estimated standard error of the model using the BPI/DECI ratio is around one-third of the size of the estimated standard errors of the model using the BPI/ECI ratio (smaller standard errors indicate a more precise estimate of the magnitude of the relationship between two variables).

As discussed in Chapter Two, differences in mission requirements and recruiting outcomes suggest that the Army's approach to recruiting may have differed in important ways from the approach taken by the other services. To consider the extent to which the inclusion of the Army in the comparisons above affect the analysis, we exclude the Army in Figure 4.16 and panel B of Table 4.1. With this sample restriction, we find an even better visual match between both ratios, with a remarkably consistent path over time between the BPI/DECI ratio and high-quality accessions for the Navy, Air Force, and Marine Corps. In terms of R-squared, the BPI/DECI ratio explains 83 percent of the variation in high-quality accessions, and the BPI/ECI ratio explains 37 percent. Results from panel C of Table 4.1, regressing Army-only high-quality accessions on each of these earnings ratios, are consistent with the notion that the Army differed in important ways from the other services, as neither model explains the pattern of high-quality Army accessions well over the sample period.

We next consider the association between the enlisted continuation rate at YOS 4 and these earnings index ratios. Figure 4.17 presents visual results for the association between enlisted continuation and the BPI/DECI and BPI/ECI ratios. Visually, one notable difference between the accession outcome considered previously and enlisted continuation is that the latter measure is much noisier, with year-over-year swings as large as 9 percentage points and multiple periods of alternating increases and decreases. Such swings could be due to changes over time in the term length of enlistment contracts, such that fewer or greater numbers of enlisted personnel sign a four-year enlistment contract and therefore are free to make a reenlistment decision at YOS 4.

Visually, the BPI/ECI ratio followed the continuation outcomes over this long period more closely, with the ECI tracking the large decline in continuation between 1996 and 1999 more closely. The regression results in panel A of Table 4.2 also show that that the ECI-based ratio performs modestly better than the DECI-based ratio in a regression of enlisted continuation on each earnings index ratio, explaining 58 percent of the variation over time, compared with 45 percent. Given our earlier point that none of these regressions controls for other factors that could affect the outcomes, it should not be surprising that we find a case where the results do not conform with expectations. In the case of continuation outcomes, these other factors could include the significant rise in civilian earnings in the mid- to late 1990s owing to the dot-com boom, which contributed to a steeper downward path for the ECI-based earnings ratio than for the DECI-based earnings ratio. This occurred just following the large post–Cold War force drawdown, which was accompanied by a financial incentive to voluntarily separate, as well as some use of involuntary separation.

To further consider how the relative strength of the statistical association may have varied around this period, we also estimated separate bivariate regression models for two, non-overlapping periods of equal length (18 years) across the full sample period: 1982 to 1999, which contains the period in question, and 2000 to 2017. We find that the overall difference in explanatory power for the regression for the full sample period is entirely based on the first of these sub-periods, during which the R-squared of the regression using the ECI-based ratio

Figure 4.16
Comparison of Enlisted DECI and ECI in Tracking the Non-Army High-Quality Accession Rate

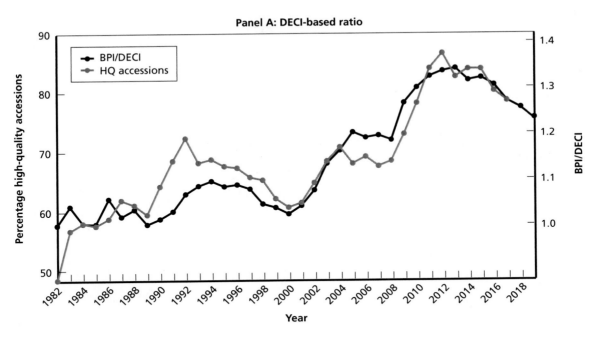

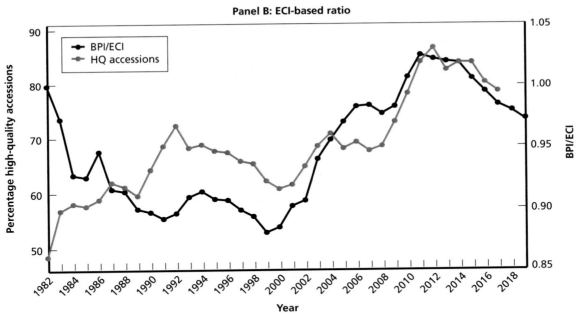

SOURCES: CPS ORG data (April–September 1982–2019) from IPUMS. ECI data from BLS. Accession data from OUSD(P&R).
NOTES: DECI cells stratified by eight age groups and four education groups. Military weights are generated from ADMF data on enlisted personnel.

Table 4.1
Regression Fit Between the High-Quality Accession Rates and Enlisted DECI Versus ECI

	Coefficient	R^2
Panel A: Percentage High-Quality Accessions—All Services		
BPI to enlisted DECI ratio	43.17***	0.493
	(7.30)	
BPI to ECI ratio	50.86*	0.107
	(24.49)	
Panel B: Percentage High-Quality Accessions—Navy, Air Force, Marine Corps		
BPI to enlisted DECI ratio	69.90***	0.830
	(5.43)	
BPI to ECI ratio	115.8***	0.365
	(26.18)	
Panel C: Percentage High-Quality Accessions—Army		
BPI to enlisted DECI ratio	2.576	0.001
	(11.82)	
BPI to ECI ratio	−56.36	0.107
	(27.91)	

NOTES: Estimates are from a bivariate regression of high-quality accession rate (in percent and as defined in each panel) on the indicated ratio (DECI or ECI). Accession data from OUSD(P&R). DECI measure uses weighting based on the composition of enlisted service members. See text and preceding figures for further details on data sources and index construction. Standard errors in parentheses. * p < 0.05, ** p < 0.01, *** p < 0.001.

is 40 percent compared to 10 percent for the DECI-based ratio. In 2000–2017 they performed about equally well (36 percent compared to 39 percent).

Figure 4.18, which plots the earnings ratios against officer continuation, shows that officer continuation at 8 YOS exhibits similar noisiness, with jumps across single years as large as 4 percentage points in magnitude. As shown in panel B of Table 4.2, both earnings ratios do a relatively poor job of explaining the variation in officer continuation, with the DECI-based ratio explaining around 10 percent and the ECI explaining 14 percent. In the same two-period analysis we described for enlisted continuation above, we find that this is due to poor performance by both measures in predicting more recent officer continuation. Prior to 2000, the R-squared is around .47 for both, but from 2000 to 2017, it is approximately zero.

Overall, these results suggest that the DECI appears to have significantly more predictive power with respect to accessions than it does for continuation. However, there are two important caveats. First, this exercise represents a simple comparison of statistical associations. Other factors, in addition to basic pay, affect recruiting and retention. Ideally, these other factors should be controlled for in isolating the effect of the respective indices on explaining recruiting

Figure 4.17
Comparison of DECI and ECI in Tracking the Enlisted Continuation Rate at 4 YOS

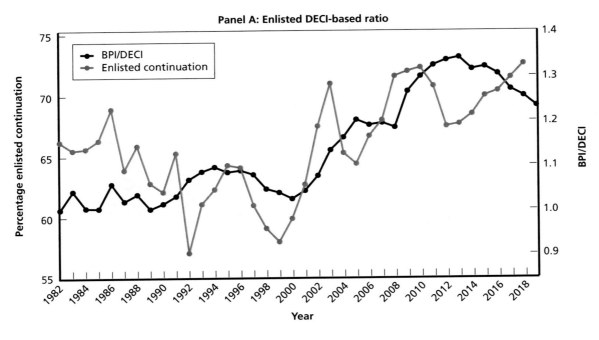

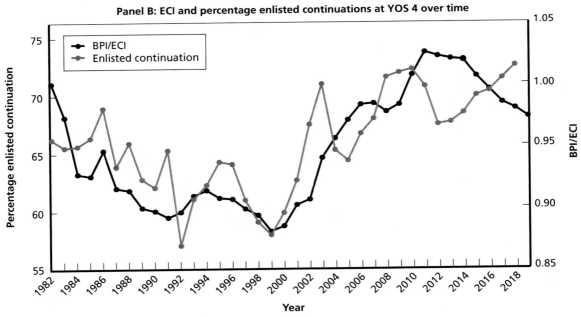

SOURCES: CPS ORG data (April–September 1982–2019) from IPUMS, ECI data from the BLS, BPI data from OUSD(P&R) (2018). Continuation data provided by DMDC.
NOTES: DECI cells stratified by eight age groups and four education groups. Military weights are generated from ADMF data on enlisted personnel.

Table 4.2
Regression Fit Between the Enlisted and Officer Continuation Rates and DECI Versus ECI

	Coefficient	R^2
Panel A: Enlisted Continuation Rate at 4 YOS		
BPI to enlisted DECI ratio	23.90***	0.449
	(4.47)	
BPI to ECI ratio	68.48***	0.582
	(9.81)	
Panel B: Officer Continuation Rate at 8 YOS		
BPI to officer DECI ratio	10.86*	0.120
	(4.96)	
BPI to ECI ratio	16.91*	0.128
	(7.45)	

NOTES: Estimates are from a bivariate regression of enlisted or officer continuation rate (%) for service members at 4 YOS on the indicated ratio. Accession data from OUSD(P&R). DECI measure uses weighting based on the composition of enlisted or commissioned officer service members. See text and accompanying figures for further details on data sources and index construction. Standard errors in parentheses. * $p < 0.05$, ** $p < 0.01$, *** $p < 0.001$.

and retention trends, Second, the performance of both the ECI and the DECI with respect to purely voluntary retention decisions, such as occurs at reenlistment, may differ meaningfully from how they perform with respect to continuation, for reasons discussed above.

Summary

The DECI, originated by RAND researchers in the early 1990s, was developed as a potential alternative to the ECI for use in guiding the annual military pay raise. The 7th QRMC considered this role for the DECI and ultimately rejected it, but the commission highlighted multiple strengths of the DECI and recommended that DoD support its further development. In this chapter, we have presented an extension and updating of the DECI, showing that it effectively addresses a number of important limitations of the ECI, which has guided the annual basic pay raise for nearly 30 years. Among the strengths of the DECI are its use of employee-based earnings data, which reflect important labor market dynamics that the employer-based earnings data used to construct the ECI are unable to capture; that it weights civilian earnings data according to characteristics of military personnel, who differ in important ways from the overall civilian labor force; that these weights are updated annually, allowing for more rapid integration of structural changes in the civilian labor market; and that the design of the DECI can accommodate a variety of potentially useful subgroup analyses, allowing it to contribute not only to the annual basic pay adjustment but also to other aspects of military compensation policy. We assessed the significant, qualitative differences in the pay guidance provided

Figure 4.18
Comparison of DECI and ECI in Tracking the Officer Continuation Rate at YOS 8

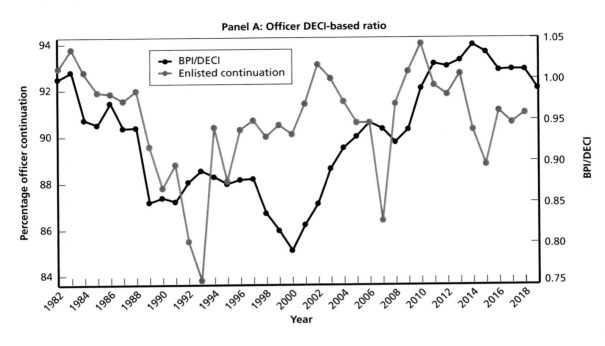

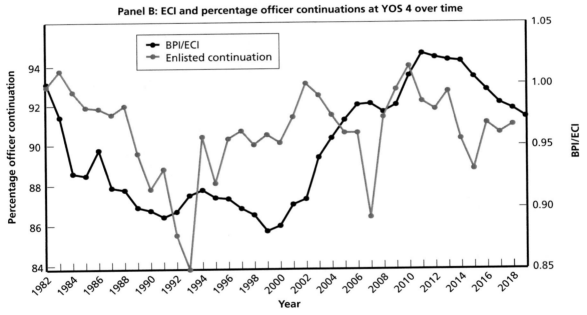

SOURCES: CPS ORG data (April–September 1982–2019) from IPUMS, ECI data from the BLS, BPI data from OUSD(P&R) (2018). Continuation data provided by DMDC.
NOTES: DECI cells stratified by eight age groups and four education groups. Military weights are generated from ADMF data on officers omitting Navy officers because of poor coding of education.

by the DECI relative to the ECI and conducted a series of empirical tests suggesting that the DECI often performs significantly better than the ECI in predicting the path of important manpower outcomes over a period of nearly 40 years.

CHAPTER FIVE

Conclusions and Policy Implications

This chapter draws from the analyses in the previous chapters to provide, first, an assessment of the continued relevance of the 70th percentile as a benchmark for setting the level of military pay, and second, an assessment of the advisability of continuing to use the ECI versus the DECI to guide the annual military pay adjustment. Specifically, we consider the following four questions in this chapter:

- What are the policy implications of the analyses?
- Has the RMC percentile needed to be higher than the 70th percentile to meet recruiting and retention objectives?
- Does the 70th percentile benchmark need to be increased?
- What are the advantages and disadvantages of the DECI versus the ECI, and are past critiques of the DECI still relevant?

We first discuss the policy implications related to the setting of the military pay level and the relevance of the 70th percentile and then discuss the potential benefits and limitations of the DECI versus the ECI. We end with our overall conclusions.

The RMC Percentile and Meeting Manpower Objectives

The weighted average RMC percentiles for enlisted personnel were 88 and 85 percent in the most recent data period, 2018–2019, using, respectively, the ADMF and the SOFS for computing earnings weights. Regardless of the data source for computing weights, these earnings percentiles far exceed the 70th percentile benchmark. They also exceed the 75 average enlisted percentile for the 1993–1997 period and the 73 percentile for 1988–1989 period that we estimate. However, we use a different methodology from the past studies that were the basis for setting the 70th percentile benchmark. The weighted average RMC percentiles for officers were 78 and 77, respectively, using the different data sources. These also exceed the average officer RMC percentile for the 1993–1997 and 1988–1989 periods that we estimate, equal to 71 for both periods. A key question is whether the RMC percentiles needed to be higher than the 70th percentile (or about the 75th percentile for enlisted and the 71st percentile for officers in our data) to meet DoD's manpower objectives.

Although DoD does not set specific objectives for retention—this is left to the individual services—it does set objectives for recruit quality, namely that 60 percent of recruits in a given service are AFQT category I–IIIA and 90 percent are Tier 1 or at least a high school graduate.

Table 5.1 replicates Table 2.1 on recruiting and retention outcomes and the factors associated with them but adds, in the first four rows, information on the weighted average RMC percentile for enlisted personnel and officers using the ADMF for weights versus the SOFS. An important conclusion is that across all services (the top panel of the table), recruit quality in the most recent period of 2018–2019 is similar to quality in the late 1980s and mid-1990s. In particular, 69 percent of recruits across DoD were in AFQT categories I–IIIA in the recent period, compared with 70 percent in the mid-1990s and 66 percent in the late 1980s. The percentage that were Tier 1 was 97 percent in the recent period, slightly higher than the percentage in the mid-1990s (95 percent) and higher than the 92 percent figure for the late 1980s. Notably, recruit quality also exceeded the benchmarks in the earlier periods.

The similarity of recruit quality in recent years with quality in the earlier periods, when quality is averaged across all of the services, could suggest, as a policy implication, that the RMC percentile for enlisted personnel needs to be higher than the 70th percentile today to sustain the recruit quality achieved when the 70th percentile was established. That is, it could imply that an RMC percentile of 70 percent is no longer relevant today and that a higher percentile for enlisted personnel is required to achieve DoD's recruit quality objectives. Some additional evidence that could support this policy implication is that the most recent period was a time of historically low unemployment, when more resources, such as higher military pay, were needed to sustain recruiting and retention. The most recent period is also one of lower recruiter productivity relative to the earlier periods, as measured by number of accessions per recruiter and higher enlistment bonus budgets (in real dollars). As shown in Table 5.1 in the top panel, recruiters achieved 11.2 annual accessions per recruiter on average in 2018–2019, compared with 15.9 on average in 1993–1997 and 22.7 in 1988–1989. One interpretation of lower recruiter productivity is that more resources—in this case recruiters—were needed to achieve an accession, meaning that a given accession was more difficult for a recruiter to achieve in recent years than in the earlier periods. If this is the case, higher military pay relative to civilian pay, as well as higher enlistment bonuses, would be needed, relative to the earlier periods, to achieve a given accession.

That said, several reasons lead us to suspect that it may be premature to assume that this policy implication is true. In other words, there is ample reason to believe that an enlisted RMC percentile as high as 88 percent is not actually needed to achieve DoD's stated manpower objectives. First, if the Army is excluded from the tabulations of outcomes and factors, as shown in the bottom panel of Table 5.1, we find that recruit quality is significantly higher in 2018–2019 relative to the earlier period, not similar. For example, the percentage of accessions that were AFQT categories I–IIIA was 75 percent, compared with 71 percent in the mid-1990s and 67 percent in the late 1980s. Furthermore, as we show in Chapter Three, we find a positive relationship between recruit quality and the weighted RMC percentile, with the association being higher when the Army is excluded. Put differently, with the exception of the Army, the services took advantage of higher military pay relative to civilian pay and a higher RMC percentile to achieve higher-quality accession cohorts. The reason for the different Army outcomes is unclear, but it appears that the Army adopted a different strategy when recruiting conditions improved at the higher RMC percentile, appearing to focus on reducing multiple recruiting-related costs while achieving its overall recruiting mission and meeting DoD recruit quality benchmarks. Knapp et al. (2018) show that the percentage of Army recruits receiving bonuses, average bonus amounts, and Army advertising expenditures dropped dramatically

Table 5.1
DoD Recruiting and Retention Outcomes and Selected Factors Related to Outcomes, 2018–2019 and Selected Benchmark Years

	Average 2018–2019	Average 2011–2013	2010	Average 1993–1997	Average 1988–1989
All Services					
Enlisted RMC percentile (ADMF weights)[a]	88	89	92	75	73
Enlisted RMC percentile (SOFS weights)[a]	85	85	86		
Officer RMC percentile (ADMF weights)[a]	78	82	82	71	71
Officer RMC percentile (SOFS weights)[a]	77	80	80		
Recruiting					
Accession mission	171,155	160,099	165,362	189,975	290,343
Percentage Tier 1	96.9	99.1	99.2	95.4	92.5
Percentage AFQT categories I–IIIA	69.0	77.0	74.4	70.4	65.5
Retention					
Enlisted continuation rate at YOS 4	73%	69%	72%	63%	62%
Enlisted continuation rate at YOS 8	85%	82%	86%	84%	87%
Officer continuation rate at YOS 8	91%	92%	94%	89%	88%
Adult unemployment rate	3.8	8.1	9.6	5.78	5.4
Military propensity	13.0	13.1	12.5	14.2	18
Enlisted end strength	1,083,131	1,134,275	1,164,553	1,277,637	1,815,034
Deployments	17,370	168,525	257,674	41,770	5,947
Recruiters	14,367	13,589	14,627	11,967	12,796
Accessions per recruiter	11.2	11.8	11.3	15.9	22.7
Enlistment bonuses ($1,000)[b]	$530,191	$343,385	$701,581	$48,357	$113,369
Reenlistment bonuses ($1,000)[b]	$1,092,816	$843,158	$1,030,551	$386,126	$865,952
All Services Except Army					
Recruiting					
Accession Mission	102,076	95,545	90,762	117,167	172,383
Percentage Tier 1	98.6	99.3	98.8	96.2	93.5
Percentage AFQT categories I–IIIA	75.0	86.2	82.5	70.9	66.9
Retention					
Enlisted continuation rate at YOS 4	76%	71%	72%	58%	64%
Enlisted continuation rate at YOS 8	85%	83%	85%	86%	87%
Officer continuation rate at YOS 8	93%	92%	93%	87%	88%
Enlisted end strength	699,828	703,085	714,233	844,998	1,156,030
Deployments	8,154	168,525	95,858	29,490	1,261
Recruiters	6,609	7,218	8,035	7,047	7,229
Accessions per recruiter	15.5	13.3	11.3	16.7	23.9
Enlistment bonuses ($1,000)[b]	$109,584	$88,327	$174,522	$19,903	$40,197
Reenlistment bonuses ($1,000)[b]	$639,936	$626,651	$778,097	$312,807	$673,932

SOURCE: OUSD(P&R) and tabulations provided by DMDC.

[a] RMC percentile for 2018–2019 is for 2018 only and is for 2012 for the period 2011–2013.

[b] Constant 2019 dollars. Enlistment bonus totals exclude Air Force.

in 2009, and the bonus budget figures in Table 5.1 imply that the Army bonus budget and number of recruiters were lower in 2018–2019 than in 2010.

Second, retention was higher in recent years relative to the 70th percentile benchmark years. As shown in Table 5.1, the enlisted YOS 4 continuation rate was substantially higher, as was the YOS 8 officer continuation rate, relative to the earlier years. The higher retention is consistent with the conclusion that the services were able to achieve better retention when the RMC percentiles for enlisted personnel and officers increased, and that higher RMC percentiles were not needed to sustain the levels of retention achieved during the 70th percentile benchmark years.

Third, even if it were the case that more resources were required in recent years to achieve the recruit quality achieved in the earlier years, it does not necessarily follow that the appropriate resource to deploy is military pay. In fact, a considerable body of research (e.g., Warner, 2010; Simon and Warner, 2007; Asch et al., 2010) finds that military pay is the least cost-effective resource for increasing high-quality recruiting and that other resources, notably recruiters and advertising, are more cost-effective. This is particularly the case when the factors that lead to recruiting difficulties are transitory, such as a particularly strong economy. This argument implies that a higher RMC percentile is not appropriate for sustaining recruiting and retention in a more difficult recruiting environment, or, alternatively, is only appropriate when other resources are found to be insufficient.

Finally, lower recruiter productivity in recent years does not necessarily mean that recruiting resources are less effective today than in the 70th percentile benchmark years. First, recruiters may be less productive simply because they are not managed as effectively as they were in the past, so they expend less effort or divert effort to less productive activities. For example, the services have moved away to some extent from recruiter incentive plans that reward individual productivity toward ones that focus on station or group output. While team-based incentives have advantages, they can also lead to free-rider problems—whereby individual team members may reduce effort when performance is not assessed at the individual level—and other issues that result in lower team output.[1] Second, if the recruiting environment were substantially more difficult for meeting accession missions, such that recruiters were less productive, we might expect enlistment propensity—the percentage of American youth reporting a positive propensity to enlist—to have fallen substantially, as well. Propensity did fall substantially relative to the late 1980s, with 13 percent of youth expressing a positive propensity in 2018 versus 22.7 in 1988–1989 (though the survey methodology for measuring propensity changed in 2001). But enlistment propensity was only somewhat lower in 2018 than in 1993–1997, when it was 14.2 percent, by about 9 percent (1 − 13.0/14.2).[2] In addition, although enlistment bonus budgets were higher in recent years than in the 70th percentile benchmark years, the size of the recruiter force and reenlistment bonus budgets—two types of resources that the services deploy when recruiting becomes more difficult—were not generally higher, when we exclude the Army from the computation.

---

[1]   See Asch (2019a) for a discussion of recruiter management, incentive plans, and recruiter productivity.

[2]   The relative stability of enlistment propensity in recent years, despite the historically low unemployment rate, could reflect the higher RMC percentile. That is, the higher RMC percentile helped sustain relative stability in enlistment propensity. The argument in the text is focused on lower recruiter productivity as an indicator of the difficulty of the recruiting market.

These arguments suggest that higher RMC percentiles—or at least an enlisted weighted percentile as high as 88 using the ADMF weights or 85 using the SOFS weights and an officer weighted percentile as high as 78 or 77, depending on data source—are not required to sustain the recruiting and retention outcomes achieved during the 70th percentile benchmark years of the late 1980s and mid-1990s. As we discussed in Chapter Two, other benchmark periods might be considered, and we showed outcomes and factors for 2010 and the period 2011–2013. But, as shown in Table 5.1 and discussed in Chapter Three, these years were ones when the RMC percentile was as high (or even higher using the ADMF weights) as it was in 2018–2019. Furthermore, recruit quality outcomes were even better, and retention was about as high. Although these other years differed in important respects, such as the number of deployments, which was substantially higher, the statistics we considered for these periods are not far off from those for the 2018–2019 period. Consequently, they do not provide additional insight into the question of whether a higher RMC percentile was required to achieve DoD manpower objectives.

Should the 70th RMC Percentile Remain the DoD Benchmark?
These findings suggest that the RMC percentiles may be too high, since recruit quality today exceeds DoD's benchmarks and, further, quality as well as both enlisted and officer retention exceed the levels observed during the late 1980s and mid-1990s, when the 70th percentile was established. That said, these findings do not necessarily imply that the 70th percentile continues to be the appropriate benchmark. In other words, it does not necessarily follow that the growth of military pay should be slowed in the future to the point that the 70th percentile benchmark is achieved, for two reasons.

First, numerous studies, including the March 2020 report of the National Commission on Military, National, and Public Service, have expressed concern about the "military-civilian divide." This divide refers to the lack of military service among most Americans; the observation that the military has become a "family business," in that recruits are more likely to have a family member who is also serving or has served; the low level of engagement between the American public and the military; and a generally shallow knowledge of the military and what service entails among American youth. For example, Asch (2019a) noted that the percentage of American youth reporting that joining the military would allow them to earn money for college fell from 85 percent in May 2004 to 60 percent in February 2016, a drop that is surprising when one recognizes that the Post-9/11 GI Bill, a program that significantly expanded education benefits for military personnel, was introduced in 2009.

While the so-called military-civilian divide has not shown up in a dramatic drop in enlistment propensity, as discussed above, this relative stability in propensity could be a result of the relatively high RMC percentile that has sustained propensity even with a more difficult recruiting market. If the military-civilian divide has made recruiting more difficult, an RMC percentile above the 70th percentile would be appropriate. Unfortunately, there have been no analyses that provide rigorous empirical evidence showing that the military-civilian divide has hurt recruiting outcomes after accounting for changes in other factors that could affect these outcomes. Nonetheless, the military-civilian divide is a concern of top military leaders, and the concern appears to be pervasive and enduring (see Ewing, 2011).

Second, the recruiting environment may be more difficult because few American youth (less than a third) would meet enlistment standards without a waiver.[3] Many disqualify for medical reasons, including weight (Orvis et al., 2018), making the rising trend in youth obesity particularly concerning. Another recent trend that challenges recruiting success is the increasing trend toward legalization of marijuana at the state and local level (Pacula and Smart, 2017; Kilmer and MacCoun, 2017). Marijuana use is still a federal offense, but some military applicants can receive a waiver if they have a history of marijuana use. Documented cases of behavioral health issues are increasing among American youth, including attention deficit hyperactivity disorder (ADHD), depression and anxiety (Danielson et al., 2018; Keyes et al., 2019). Finally, waivers are not allowed for some behavioral health issues, such as self-harm.

These arguments suggest a more difficult recruiting environment than in earlier periods due to factors that are not transitory (such as a historically low unemployment rate). Together, they suggest that an RMC percentile benchmark above the 70th is appropriate. But if the 85th percentile for enlisted personnel and the 77th percentile for officers are too high (as measured using the SOFS weights) and the 70th too low, what is the right number? We do not have a specific number to offer, but we believe a figure of around the 75th to 80th percentile for enlisted personnel and around the 75th percentile for officers are likely to be appropriate. Given the evidence that increasing military pay is associated with improved recruit quality and retention—even holding constant other factors, including other resources such as bonuses, advertising, and recruiters—increasing the benchmark would address a more challenging environment and help ensure that the services could continue to meet and even exceed the DoD benchmarks for recruit quality and achieve the retention levels for enlisted personnel and officers that were achieved when the 70th percentile benchmark was set.

An Important Caveat: Increased Requirement for More High-Quality Recruits

The analysis shows that the services, except for the Army, increased recruit quality when the RMC percentile increased, and retention and the experience level of the enlisted and officer force improved, as well. Such increases improve defense capability, and, in the absence of empirical evidence to the contrary, it might have been optimal for the Navy, Air Force, and Marine Corps to increase their percentages of high-quality recruits and to increase retention. Changes in defense threats, readiness requirements, and military technology have shifted manpower requirements toward higher-AFQT recruits in some services. For example, the Navy's weapon systems have become more complex in the past decade. Its widened use of software-based technology is needed to support network-centric warfare, causing an increase in the demand for personnel with information technology skills and the ability to handle complex information in decisionmaking (Wenger, Miller, and Sayala, 2010).

If the services now require a greater proportion of high-quality recruits to meet changing readiness requirements, the DoD benchmarks for recruit quality should reflect that increased requirement. The current recruit quality benchmarks were set based on considerable research using data from the 1980s, such as the work of "Project A" discussed in Chapter One. Any changes in the benchmark should similarly be guided by research using more recent data. Put differently, DoD and the services should validate any changes in recruit quality requirement as well as requirements regarding retention and should provide rigorous analysis demonstrating

---

[3]   The estimate of youth qualification for enlistment comes from National Commission on Military, National, and Public Service (2020).

that higher quality benchmarks and a more experienced force are required to achieve today's readiness objectives.

Potential Benefits and Limitations of the DECI

Although the ECI has multiple appealing qualities—most notably, timeliness, representativeness, and accuracy—that have led to its longstanding use in guiding the annual adjustment in military pay, it also has important limitations that may limit its effectiveness from the perspective of providing information on the civilian labor market opportunities *relevant to active duty military personnel*. Multiple aspects of the DECI—the use of underlying data that reflect labor market outcomes from an employee, rather than an employer, perspective; weighting that reflects the age and educational attainment composition of military personnel; and the flexibility to assess paths of earnings growth for specific subgroups of interest within the overall active duty force—make the DECI a promising alternative, or supplement, to the ECI. We conclude here with a brief discussion of the pros and cons of the DECI relative to the ECI, focusing on five key issues: accuracy, timeliness, cost, flexibility, and relevance.

Accuracy of the DECI

Two dimensions of accuracy are relevant. The first is the accuracy of the CPS ORG earnings data. The CPS began as an independent survey of unemployment under the auspices of the Works Progress Administration during the Great Depression before being transferred to the U.S. Census Bureau in the early 1940s (U.S. Census Bureau, 2006). As the source of official national measures of unemployment and labor force participation, as well as an important source of information on national demographics, population mobility, earnings, and other key measures of social and economic change, the CPS has an outsized role in policymaking. Consequently, the methodology, implementation, and accuracy of the CPS are regularly scrutinized and updated, making it arguably the most accurate source of data for computing U.S. labor force statistics.

The second is the accuracy of the military data used to weight the CPS earnings data. The DECI utilizes end-of-fiscal-year (September) administrative data on all active duty personnel contained in the DMDC ADMF. Over the nearly four-decade time span we consider, we found modest data quality issues, primarily related to the accuracy and completeness of the coding of educational attainment and to the issue of missing age data in a handful of years,[4] but overall these issues were minimal. The greatest incidence of missing education data across DoD, in 2018, was 1.27 percent, and the average over all years was around 0.5 percent. The SOFS could also be effectively used as a source of weights in the future (should the DECI be implemented), but it relies on voluntary sampling of active duty personnel, so it is not fully representative of the exact composition of military personnel. The ADMF has the advantage of representing not a sample but the actual population of active duty military personnel in each year. Additionally, our comparison of a DECI constructed using SOFS and ADMF data revealed little difference in their paths across a period of 16 years.

---

[4]　In our analysis, we successfully imputed age from surrounding years for the vast majority of those missing age data, which was confined to the years 1995 through 1997.

Timeliness of the DECI

At the time that the Hosek 1992 study was published, data access was a significant hurdle for all kind of quantitative analyses, even for a ubiquitous survey such as the CPS. Nearly three decades on, the issue of timely access to data such as the CPS has been resolved by advances in computing power and connectivity. Not only does the BLS typically make CPS micro-data available through its File Transfer Protocol (FTP) site within a week of the completion of monthly surveys, but these data are then quickly integrated into a user-friendly CPS data repository, the Integrated Public Use Microdata Series (IPUMS; Flood et al., 2020). IPUMS is run by the University of Minnesota and supports the generation of a customized set of CPS data for any number of years over nearly half a century. Furthermore, many variables have been harmonized across time and come with clear and well-organized documentation. The IPUMS system also maintains a record of all requests from each data user, so that pulling new data requires nothing more than going to a prior data request, changing the sample to a current month, and sending the revised data request. These data are delivered over the internet along with multiple data dictionaries compatible with most common statistical software packages. IPUMS data are typically ready for use 7–10 days after the data are first made available by the BLS, so data are available through this conduit around three weeks following the end of the relevant survey month.

Because of these changes, it is now possible to use monthly CPS data collected through the end of September to generate a DECI (or multiple DECIs) by late October. This advance in data availability eliminates any meaningful difference in the timeliness of the DECI relative to the ECI.

Cost

The ECI has the advantage of being generated using the time, resources, and expertise of another federal agency. However, generating the DECI is readily within the grasp of OSD. The software code written to generate the DECIs for this study is provided in Appendix E. Using this code, updated DECIs can be generated using the in-house computing resources and expertise found in organizations such as the Office of People Analytics in OUSD(P&R) or the Office of Cost Assessment and Program Evaluation within DoD. The Office of Cost Assessment and Program Evaluation provides independent analytics for the Secretary of Defense, so computation of the DECI by this office could help ensure that the computation is performed independent of OUSD(P&R). The code provided in Appendix E could further be used to assist with a periodic audit of the DECI to ensure its continued accuracy.

Furthermore, relative to a study like Hosek 1992 conducted in the early 1990s, advances in computer technology have dramatically reduced both the financial and resource opportunity cost of generating a custom index for the sole purpose of informing the annual military pay adjustment. Based on the work in this study, we estimate that, once the code has been implemented and the administrative data prepared, the data work for computing the DECI annually could be run in less than an hour on a modern laptop computer.

Flexibility

The fact that a DECI can be created for any military subgroup of interest increases its potential usefulness to policymakers concerned with setting and adjusting military basic pay. As we show in Appendix D, subgroup DECIs can be constructed for officers and enlisted personnel, providing information on earnings growth for civilians more specifically relevant to these groups that differ meaningfully from one another in terms of age and educational attainment. Subgroup DECIs could also be generated by a given service to inform the process of setting special and incentive pays in order to flexibly address potential retention issues related to specific differences in the civilian earnings opportunities of personnel in any number of pay grade or occupational subgroups. As an example, a DECI could be constructed for civilians working in cyber-related occupations to provide information relative to the recruitment and retention of military personnel in cyber-related career fields. Note that, though we have used only age, educational attainment, and gender in the present analysis, the DECI framework can easily incorporate occupation-specific earnings data in cases when these occupational linkages are well defined and germane to the policy question.

Relevance

Because the DECI uses weights that reflect the demographics of military personnel, it captures civilian earnings growth more relevant to military personnel. In particular, three out of four service members exiting the military do not have a prearranged job, implying that most members typically enter the broader labor market rather than a specific occupational field (Parker et al., 2019). Consequently, their assessment of their civilian opportunities will be based on their level of education, experience (as proxied for by age in our analysis), and other characteristics. Data on employer costs of compensation that do not account for these characteristics, such as the National Compensation Survey data used to generate the ECI, will be less relevant. Related to this point, changes in macroeconomic conditions may affect the level of education and experience required by new employers for a given job—so-called "upskilling" and "downskilling" (Sasser, Shoag, and Ballance, 2016, 2019)—and these employment dynamics cannot be captured using a job-based employment sample, since the data do not contain detailed demographic information on the job holders. CPS data easily capture "upskilling" and "downskilling," since the CPS captures a sample of employees, rather than jobs, and collects information on age, educational attainment, gender, and many other demographic characteristics.

The DECI is also more relevant to the setting of the pay adjustment for military personnel because it is more sensitive to macroeconomic fluctuations than the ECI. A lack of sensitivity to fluctuations in civilian earnings based on macroeconomic shocks was portrayed by the 7th QRMC as a virtue of the ECI and a potential drawback of the DECI. However, as discussed earlier in the chapter, the ECI tends to mute, reduce, or even negate civilian pay growth during recessions, thereby muting the effect of these changes in average civilian earnings on the suggested annual military pay adjustment. Using the ECI helps to limit any negative impact of recessions on the pay raise for military personnel, but using an index that is insensitive to significant changes in the civilian labor market to set the annual pay raise is not a cost-effective policy from the standpoint of the taxpayer. During recessions, recruiting and retention tend to improve, implying that pay raises should be more limited during recessions than the ECI might suggest. By being more sensitive to macroeconomic fluctuations, the DECI would provide more accurate and relevant guidance for sustaining recruiting and retention at lower cost.

Finally, as shown by the exercises in the previous subsection, the changes in civilian earnings measured by the DECI impart more relevant information with respect to accession and continuation outcomes.

Wrap Up

We computed weighted RMC percentiles for enlisted personnel and officers adjusting for the education distribution of military personnel. The RMC percentile today far exceeds the 70th percentile benchmark set based on data from the late 1980s and mid-1990s, reflecting the relatively fast military pay growth from the late 1990s to 2010, as well as a downward trend in real civilian wages. Similar patterns emerge in our analyses by age group. The average RMC percentile is strongly correlated with recruit quality through time, and, importantly, the higher RMC percentile observed in the 2018–2019 period is associated with higher recruit quality in all but the Army relative to the late 1980s and mid-1990s. Our analysis argues for a new RMC benchmark that is less than the high levels seen today to achieve current recruit quality objectives. We argue that the RMC percentile benchmark should exceed the 70th percentile, given concerns about the military-civilian divide and the ability of American youth to meet enlistment standards. Although we do not provide a specific figure for the new benchmark, a percentile between the 75th and the 80th percentile (using the SOFS weights) seems reasonable.

With respect to the ECI versus the DECI, the analysis points to several key benefits of the DECI that directly address shortcomings of using the ECI as the key indicator informing the annual adjustment in military pay. Conceptually, the DECI is built to measure the path of earnings growth for a population of civilian workers that reflects the composition of the military at any given point in time. The paradox of the "pay gap" of the 1980s and 1990s when civilian earnings were measured by the ECI essentially disappeared when the DECI was used in its place. We also find that the DECI performs as well or sometimes substantially better than the ECI in tracking key manpower outcomes. Additionally, the flexibility of the DECI means that it can provide important information about key subgroups, such as cyber personnel and officers versus enlisted personnel. Finally, because of significant changes in both data availability and computing power, the DECI can now be generated in a timely fashion using resources and expertise already available within DoD and can be periodically audited using the computer code in Appendix E as the standard to ensure its continued accuracy.

Trends over Time, by Service

This appendix shows continuation rates and various factors associated with recruiting and retention outcomes, including enlisted continuation rates at 4 and 8 YOS, officer continuation rates at 8 YOS, high-quality enlisted DoD accessions and the adult unemployment rate, the number of enlisted personnel receiving Imminent Danger Pay or Hostile Fire Pay relative to the number receiving these pays in the year 2000, the number of recruiters in each service relative to the number in the year 1989, and the annual number of accessions per recruiter.

Figure A.1
Continuation Rate of Enlisted Personnel at YOS 4, by Service

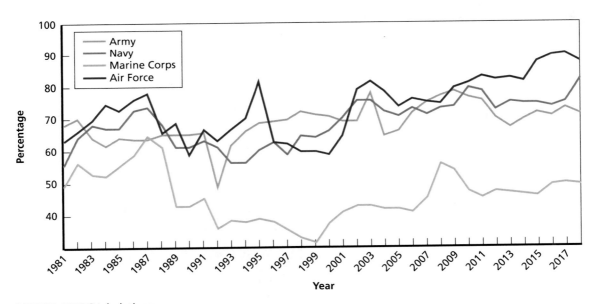

SOURCE: DMDC tabulations.
NOTE: A "continuer" is an active duty member at YOS 4 at the beginning of the year who has not changed service and remains in the ADMF at the start and end of the year.

Figure A.2
Continuation Rate of Enlisted Personnel at YOS 8, by Service

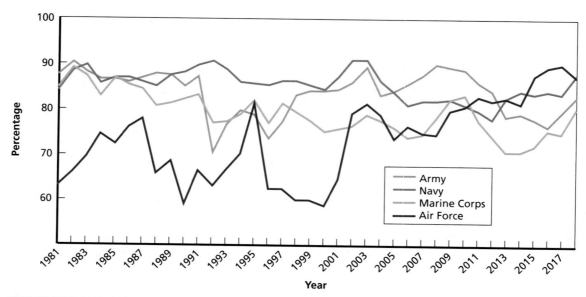

SOURCE: DMDC tabulations.
NOTE: A "continuer" is an active duty member at YOS 8 at the beginning of the year who has not changed service and remains in the ADMF at the start and end of the year.

Figure A.3
Continuation Rate of Officers at YOS 8, by Service

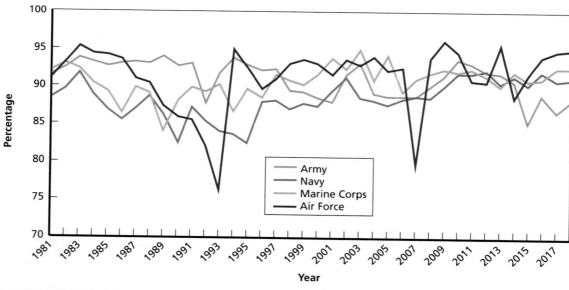

SOURCE: DMDC tabulations.
NOTE: Tabulations are not available for DoD overall. A "continuer" is an active duty member at YOS 8 at the beginning of the year who has not changed service and remains in the ADMF at the start and end of the year.

Figure A.4
High-Quality DoD Enlisted Accessions and Adult Unemployment Rate

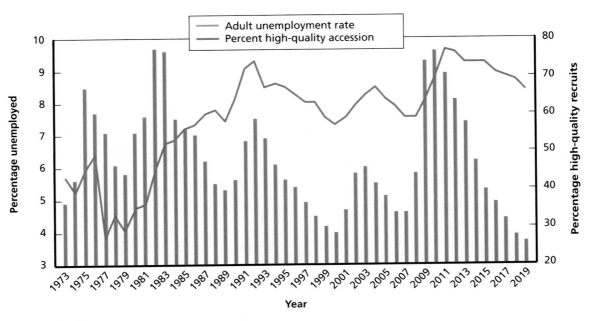

SOURCE: Pop Rep.

Figure A.5
Enlisted Personnel Receiving Imminent Danger or Hostile Fire Pay

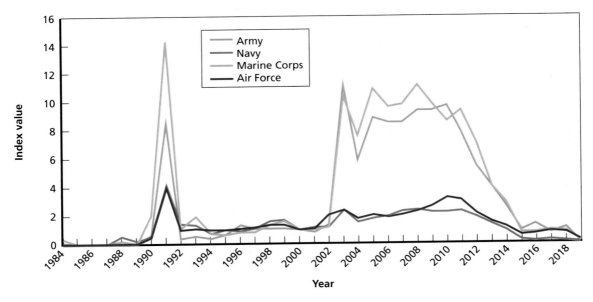

SOURCE: DMDC tabulations.
NOTE: The year 2000 = 1.00.

Figure A.6
Index of Annual Average Monthly Recruiter Count, by Service

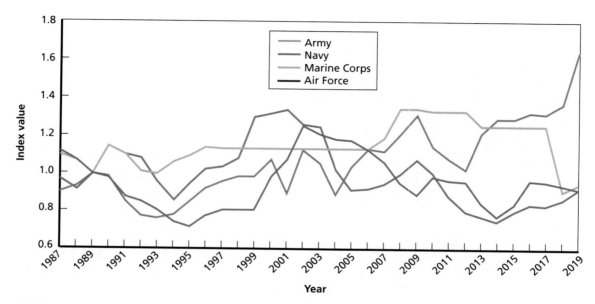

SOURCE: OUSD(P&R), Office of Accession Policy.
NOTE: The year 1989 = 1.00

Figure A.7
Annual Accessions per Recruiter, by Service

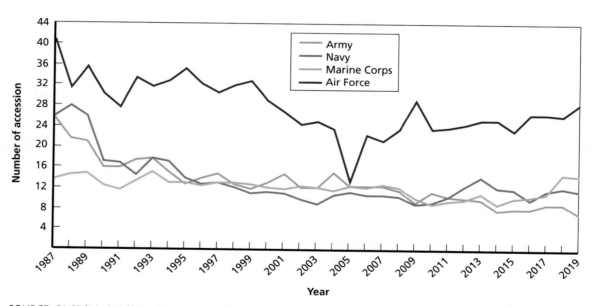

SOURCE: OUSD(P&R), Office of Accession Policy.

Average Regular Military Compensation, by Rank and Years of Service

In this appendix, we present graphs that compare RMC to civilian wages from 1975 to 2018. These graphs are similar to those in the second section of Chapter Three but use average RMC, as reported in the Greenbooks, instead of median RMC as calculated from the Active Duty Pay files. Here we also focus on a specific grade, E-5 for enlisted and O-3 for officers, and those with 6 or 7 YOS instead of segmenting the data by age as we do in Chapter Three.

In comparing Figures B.3 and B.4 for O-3s to the graphs for officers in Chapter Three, Figures 3.7 and 3.8, we see that the pattern of how RMC percentiles evolved through time is similar despite the specific population of interest, although the percentiles for O-3s are in general a bit higher than those for the age categories chosen in Chapter Three. However, in comparing Figures B.1 and B.2 to the graphs for enlisted in Chapter Three, Figures 3.5 and 3.6, we see that, although the patterns through time are similar, E-5s have much higher RMC percentiles than those shown for the given enlisted age categories in Chapter Three. Wages for E-5s also appear to have continued to increase between 2011 and 2017, while they remained flat for enlisted personnel in age categories 23–27 and 28–32 in Figures 3.5 and 3.6. Some of these differences between the two sets of graphs are attributable to our use of different data sources for RMC (Active Duty Pay files versus Greenbooks) or the fact that we are comparing median RMC in Chapter Three versus average RMC here. However, the majority of the difference comes from the different mix of individuals included in the two populations. While we don't have the exact underlying data of the observations that were used in the Greenbook calculations, we can get some sense of how the two populations compare using the ADMF data. For example, while the majority of E-5s with 6 or 7 YOS are between the ages of 23–32, only about 60 percent of enlisted in that age group with 6 or 7 YOS are E-5s and over 27 percent are E-4s, which have a lower wage. Across all years of service, only about 37 percent of enlisted personnel between the ages of 23–32 are E-5s, and more the 40 percent are E-3s and E-4s. Thus, when dividing the data by age category, as we do in Chapter Three, RMC percentiles are lower than when looking only at E-5s, as we do in this Appendix.

The differences in measured RMC percentiles of the civilian wage distribution between Figures B.1 and B.2 in this appendix and 3.5 and 3.6 in Chapter Three demonstrate the sensitivity of these analyses to sample construction and the importance of choosing an appropriate comparison group in evaluating the adequacy of RMC. Because results differ with different comparison groups, we prefer the weighted average RMC analyses at the beginning of Chapter Three, as they explicitly adjust for differences in wages by education and YOS.

Figure B.1

Civilian Wages for High School Graduate Men with 6 or 7 YOS and Average Regular Military Compensation for an E-5 with 6 or 7 YOS, Calendar Years 1975–2018, in 2019 Dollars

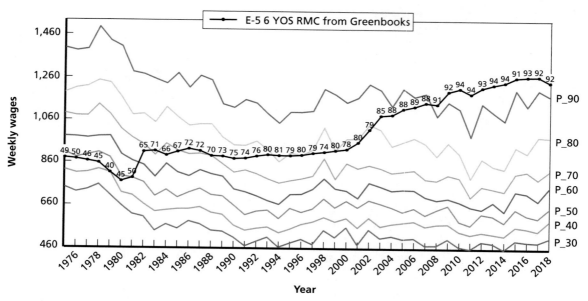

SOURCES: Greenbooks, 1975–2018; CPS, 1981–2019.
NOTES: The reference population is men with 6 or 7 YOS who reported high school completion as their highest level of education, worked more than 35 weeks in the year, and usually worked more than 35 hours per week. We computed the weekly wage by dividing annual earnings by annual weeks worked. The colored lines depict the wages at the indicated percentiles (on the right axis) of the wage distribution for this population. For instance, at the 70th percentile, 30 percent of the population had higher wages and 70 percent had lower wages. The black line depicts average RMC for an E-5 with 6 or 7 YOS from the Greenbooks. The numbers above the RMC line are the percentiles at which RMC stood in the population's wage distribution.

Figure B.2
Civilian Wages for Men with Some College with 6 or 7 YOS and Average Regular Military Compensation for an E-5 with 6 or 7 YOS, Calendar Years 1975–2018, in 2019 Dollars

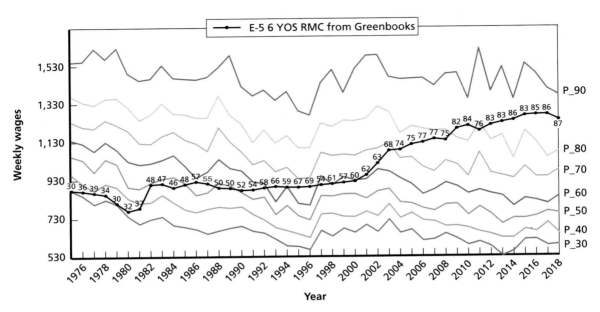

SOURCES: Greenbooks, 1975–2018; CPS, 1981–2019.
NOTES: The reference population is men with 6 or 7 YOS who reported some college as their highest level of education, worked more than 35 weeks in the year, and usually worked more than 35 hours per week. We computed the weekly wage by dividing annual earnings by annual weeks worked. The colored lines depict the wages at the indicated percentiles (on the right axis) of the wage distribution for this population. For instance, at the 70th percentile, 30 percent of the population had higher wages and 70 percent had lower wages. The black line depicts average RMC for an E-5 with 6 or 7 YOS from the Greenbooks. The numbers above the RMC line are the percentiles at which RMC stood in the population's wage distribution.

Figure B.3
Civilian Wages for Men with a Bachelor's Degree with 6 or 7 YOS and Average Regular Military Compensation for an O-3 with 6 or 7 YOS, Calendar Years 1975–2018, in 2019 Dollars

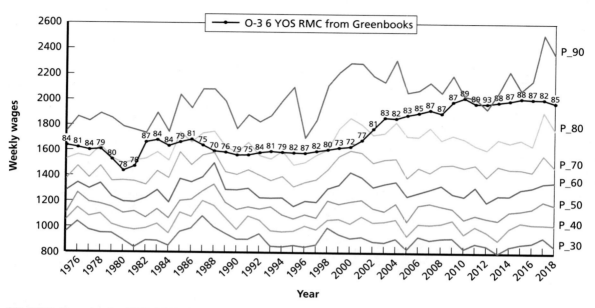

SOURCES: Greenbooks, 1975–2018; CPS, 1981–2019.
NOTES: The reference population is men with 6 or 7 YOS who reported a Bachelor's degree as their highest level of education, worked more than 35 weeks in the year, and usually worked more than 35 hours per week. We computed the weekly wage by dividing annual earnings by annual weeks worked. The colored lines depict the wages at the indicated percentiles (on the right axis) of the wage distribution for this population. For instance, at the 70th percentile, 30 percent of the population had higher wages and 70 percent had lower wages. The black line depicts average RMC for an O-3 with 6 or 7 YOS from the Greenbooks. The numbers above the RMC line are the percentiles at which RMC stood in the population's wage distribution.

Figure B.4
Civilian Wages for Men with a Master's Degree or more with 6 or 7 YOS and Average Regular Military Compensation for an O-3 with 6 or 7 YOS, Calendar Years 1975–2018, in 2019 Dollars

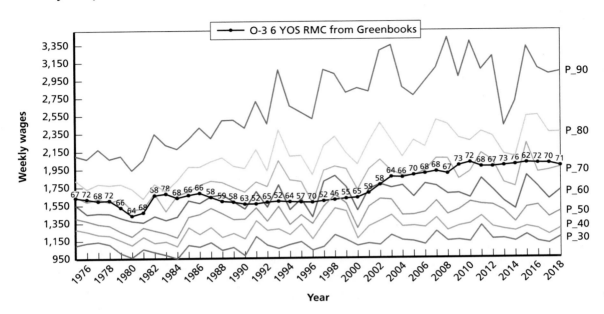

SOURCES: Greenbooks, 1975–2018; CPS, 1981–2019.
NOTES: The reference population is men with 6 or 7 YOS who reported a Master's degree or more as their highest level of education, worked more than 35 weeks in the year, and usually worked more than 35 hours per week. We computed the weekly wage by dividing annual earnings by annual weeks worked. The colored lines depict the wages at the indicated percentiles (on the right axis) of the wage distribution for this population. For instance, at the 70th percentile, 30 percent of the population had higher wages and 70 percent had lower wages. The black line depicts average RMC for an O-3 with 6 or 7 YOS from the Greenbooks. The numbers above the RMC line are the percentiles at which RMC stood in the population's wage distribution.

Process for Estimating Weighted Average Regular Military Compensation

In this appendix, we outline the process and necessary files for estimating weighted average RMC using the SOFS education distribution. Relevant code for reading in the CPS data in SAS, as well as code to combine the various datasets and perform the weighted average calculations in STATA, is also included. Note that users will need to change the directory for the programs to reflect where the data are stored on their system. We have used @input_directory and @output_directory to indicate where these changes in the code need to occur.

Necessary Data

Data from the following sources are necessary to perform the weighted average calculation:

Current Population Survey (CPS) March Supplement: We take the data from the National Bureau of Economic Research (NBER) data archives found at http://data.nber.org/data/current-population-survey-data.html. Note that the year listed corresponds to the release year, but that the data are gathered the previous year. Thus, the March Supplement 2019 data, which were released around September 2019, are from calendar year 2018, and we make this adjustment in comparing with the military pay data. On the NBER CPS Supplements webpage, data and documentation files can be found by looking under the "Mar" column for the given year. Clicking on the relevant data or documentation icon will download the files locally. For 2019, the following files are used in the first SAS program below: "ffpub19.csv," "hhpub19.csv," and "pppub19.csv."

The OSD Directorate of Compensation's *Selected Military Compensation Tables* (OUSD[P&R], Directorate of Compensation, 1980–2018), also known as the **Greenbook**: We use data from the Greenbooks on average salaries by rank and YOS as well as number of active duty personnel by rank and YOS. The former comes from the table titled "Detailed RMC Tables for All Personnel: Assume All Cash Pay," which in recent years is Table B4, and the latter comes from "Military Personnel by Pay Cell," which is Table A6. We use data from all YOS for E-1 through E-9 and O-1 through O-6 as well as the "ALLCO" and "ALLENL" rows in both tables.

We use data compiled for us by DoD's Office of People Analytics from the **Status of Forces Surveys (SOFS)** on the educational distribution of enlisted and officers by rank. For E-1 through E-9, SOFS data provide the percentage of personnel that fall in each of the following categories: "Non High School Graduate," "High School Graduate," "Less Than 1 Year of College," "1 or More Years of College (No Degree)," "Associate's Degree," "Bachelor's Degree,"

"Master's/Doctoral/Professional School Degree." For O-1 through O-6, they provide the percentage of personnel that fall in each of the following categories: "High School Graduate," "Less Than 1 Year of College," "Some College," "College Graduate or More," "Advanced Degree." Note that "College Graduate or More" here includes those with an associate's degree although they make up only a very small percentage of officers.

For estimates of the gender mix of military members, we use ***Population Representation in the Military Services: Fiscal Year 2018*** (OUSD[P&R], 2020), also known as "Pop Rep," which can be found at https://www.cna.org/research/pop-rep. From this document, we use the total DoD percent column from "Table D-13. Female Active Component Enlisted Members by Service with Civilian Comparison Group, FYs 1970–2018" and the total DoD percent column in the second part ("Corps" rather than "Gains") from "Table D-19. Female Active Component Commissioned Officer Gains and Corps by Service, FYs 1973–2018."

Finally, to adjust for inflation to current dollar amounts we use "All items in U.S. city average, all urban consumers, not seasonally adjusted," from the **Bureau of Labor Statistics, CPI for All Urban Consumers (CPI-U)**, Series Id:CUUR0000SA0, found at https://data.bls.gov/timeseries/cuur0000sa0?series_id=cwur0000s.

The code below is organized into three sections. The first section (three programs) is SAS code to read in and process the CPS data and to make the data ready to use for the subsequent programs. The second section of code (two programs) is in Stata, and it computes education distributions for enlisted and officers from the SOFS. The third section of code (two programs) is in Stata, and it creates the final weighted average RMC graphs from Chapter Three.

Code to Read-in and Process the CPS Data

We first read in the CPS data for the current year and then add it to previous years' CPS data using SAS.

Read in 2019 CPS Data (Corresponding to Calendar Year 2018)
Input: file downloaded from NBER website
(http://www.nber.org/data/current-population-survey-data.html)
Output: cpsmar2019select
Program name: "read_cpsmar19_select.sas"

```
*Set library names;
libname IPUMS "@input _ directory";
libname library "@input _ directory";
options nocenter linesize=256;


*The labels on this data are 2019, but it represents the 2018 census data;


*Pulling in only the variables from the CPS data that are necessary;
```

```
*links between tables;
*hh            ff            pp
 h _ seq   =   fh _ seq   =   ph _ seq
               ffpos      =   phf _ seq
;
*****HH Fields*****;
               /*
               H _ SEQ
               Household sequence number
               5 29 (00001:99999)
               Values: 00001- 99999=Household sequence number
               Universe: All Households
               **This is unique in hhpub
               */
*****FF Fields***;
               /*
               FH _ SEQ
               Household sequence number. Matches H _ SEQ for samehousehold
               5 4 (00001:99999)
               Values: 00001-99999 = household
               */
               /*
               FFPOS
               Unique family identifier. This field plus FH _ SEQ results in a
               unique family number for the file.
               2 2 (01:16)
               Values: 01-39 = index for family identifier
               Universe: All Families
               */
****PP Fields*****;
               /*
               PH _ SEQ
               Household seq number
               5 36 (00000:99999)
               Values: 00001:99999
               Universe: All Persons
               */
               /*
               PHF _ SEQ
               Pointer to the sequence number of own family record in household.
               (Care should be exercised when using these data as the related
               subfamilies are a part of the primary family and usually their
               characteristics come from the primary family record)
               2 41 (01:16)
               Values: 01:16
               Universe: All Persons
```

```
                        */
    *Importing the data;
    proc import datafile="@input_directory/hhpub19.csv"
                 out=hhpub
                 dbms=csv
                 replace;
                 getnames=yes;
    run;


    proc import datafile="@input_directory/ffpub19.csv"
                 out=ffpub
                 dbms=csv
                 replace;
                 getnames=yes;
    run;


    proc import datafile="@input_directory/pppub19.csv"
                 out=pppub
                 dbms=csv
                 replace;
                 getnames=yes;
    run;


    ****************************************************************;
    *Create file structure;
    *Attach household and family data;
    data hhpub; set hhpub; record_hhpub=1; run;
    data ffpub; set ffpub; record_ffpub=1; run;
    data pppub; set pppub; record_pppub=1; run;
    proc sql;
    create table structure as
    select a.h_seq,a.record_hhpub,a.gestfips,
           b.fh_seq,b.ffpos,b.record_ffpub,
           c.ph_seq,c.phf_seq,c.record_pppub,
                    c.p_stat,c.a_age,c.a_hga,c.prdtrace,c.prdthsp,c.wkswork,c.
           hrswk,c.wsal_val,
           c.a_sex,c.marsupwt,c.a_fnlwgt,c.wsal_yn
    from hhpub as a full outer join ffpub as b
                    on a.h_seq=b.fh_seq
                  full outer join pppub as c
              on b.fh_seq=c.ph_seq and b.ffpos=c.phf_seq;
    quit;


    *Label data;
    proc format  library=ipums;
```

```
*Note that default=32 is the default length of the format (not the value of the
variable);
VALUE a _ age        (default=32)
        80        =   "80-84 years of age"
        85        =   "85+ years of age"
;
VALUE a _ sex        (default=32)
        1         =   "Male"
        2         =   "Female"
;
VALUE a _ hga        (default=32)
        0         =   "Children"
        31        =   "Less than 1st grade"
        32        =   "1st,2nd,3rd,or 4th grade"
        33        =   "5th or 6th grade"
        34        =   "7th and 8th grade"
        35        =   "9th grade"
        36        =   "10th grade"
        37        =   "11th grade"
        38        =   "12th grade no diploma"
        39        =   "High school graduate - high school dip"
        40        =   "Some college but no degree"
        41        =   "Associate degree in college - occupati"
        42        =   "Associate degree in college - academic"
        43        =   "Bachelor's degree (for example: BA,AB,"
        44        =   "Master's degree (for example: MA,MS,ME"
        45        =   "Professional school degree (for exampl"
        46        =   "Doctorate degree (for example: PHD,EDD"
;
VALUE prdtrace       (default=32)
        1         =   "White only"
        2         =   "Black only"
        3         =   "American Indian,Alaskan Native only (A"
        4         =   "Asian only"
        5         =   "Hawaiian/Pacific Islander only (HP)"
        6         =   "White-Black"
        7         =   "White-AI"
        8         =   "White-Asian"
        9         =   "White-HP"
        10        =   "Black-AI"
        11        =   "Black-Asian"
        12        =   "Black-HP"
        13        =   "AI-Asian"
        14        =   "AI-HP"
        15        =   "Asian-HP"
        16        =   "White-Black-AI"
```

```
        17        =   "White-Black-Asian"
        18        =   "White-Black-HP"
        19        =   "White-AI-Asian"
        20        =   "White-AI-HP"
        21        =   "White-Asian-HP"
        22        =   "Black-AI-Asian"
        23        =   "White-Black-AI-Asian"
        24        =   "White-AI-Asian-HP"
        25        =   "Other 3 race comb."
        26        =   "Other 4 or 5 race comb."
    ;
    VALUE p _ stat       (default=32)
        1         =   "Civilian 15+"
        2         =   "Armed Forces"
        3         =   "Children 0 - 14"
    ;
    VALUE prdthsp        (default=32)
        0         =   "Not in universe"
        1         =   "Mexican"
        2         =   "Puerto Rican"
        3         =   "Cuban"
        4         =   "Dominican"
        5         =   "Salvadoran"
        6         =   "Central American, (exc. Salv)"
        7         =   "South American"
        8         =   "Other Hispanic"
    ;
    VALUE wkswork        (default=32)
        0         =   "Not in universe"
        1         =   "1 week"
        52        =   "52 weeks"
    ;
    VALUE hrswk          (default=32)
        0         =   "Not in universe"
        1         =   "1 hour"
        99        =   "99 hours plus"
    ;
    VALUE wsal _ val     (default=32)
        0         =   "None or not in universe"
    ;
    VALUE a _ fnlwgt     (default=32)
        0         =   "Supplemental Spanish sample"
    ;
    VALUE wsal _ yn      (default=32)
        0         =   "Not in universe"
        1         =   "Yes"
```

```
        2              =    "No"
;
run;

data ipums.cpsmar2019select;
format
        a _ age a _ age.
        a _ sex a _ sex.
        a _ hga a _ hga.
        prdtrace        prdtrace.
        p _ stat        p _ stat.
        prdthsp         prdthsp.
        wkswork         wkswork.
        hrswk  hrswk.
        wsal _ val      wsal _ val.
        a _ fnlwgt      a _ fnlwgt.
        wsal _ yn wsal _ yn.;

set structure;

a _ fnlwgt=a _ fnlwgt/100;
marsupwt=marsupwt/100;

attrib  a _ age         length=3        label="Item 18d - Age";
attrib  a _ fnlwgt      length=8        label="Final weight (2 implied decimal
                                        places";

attrib  a _ hga         length=3        label="Item 18h - Educational attainment";
attrib  a _ sex         length=3        label="Item 18g - Sex";
attrib  ffpos           length=3        label="Unique family identifier";
attrib  fh _ seq        length=4        label="Household sequence number";
attrib  gestfips        length=3        label="State FIPS code";
attrib  hrswk           length=3        label="Item 41 - In the weeks that ...
                                        worked";

attrib  h _ seq         length=4        label="Household sequence number";
attrib  marsupwt        length=8        label="Supplement final weight (2 implied";
attrib  phf _ seq       length=3        label="Pointer to the sequence number of
                                        own";

attrib  ph _ seq        length=4        label="Household seq number";
attrib  prdthsp         length=3        label="Detailed Hispanic recode";
attrib  prdtrace        length=3        label="Race";
attrib  p _ stat        length=3        label="Status of person identifier";
attrib  wkswork         length=3        label="Item 33 - During 20.. in how many
                                        week";

attrib  wsal _ val      length=5        label="Recode - Total wage and salary
                                        earning";

attrib  wsal _ yn       length=3        label="Recode";
```

```
run;
proc contents data=ipums.cpsmar2019select;
run;
```

Combine 2019 CPS Data with Previous CPS Data

Inputs: cpsmar99 – cpsmar12,cpsmar2013, cpsmar2014t, cpsmar2015 – cpsmar2018,
cpsmar2019select
Output: cps9919
Program name: "stack_cpsmar19.sas"

```
libname jum        "@input _ directory";
libname cpssas "@input _ directory";

options nocenter linesize=256;

proc format cntlin=jum.fcpsmar2018;
run;


*The older cps data has state coded differently;
*http://ceprdata.org/wp-content/cps/CPS _ March _ Codebook _ 1999.pdf;
proc format;
value newstate
63='1'  /*Alabama              */
94='2'  /*Alaska               */
86='4'  /*Arizona              */
71='5'  /*Arkansas             */
93='6'  /*California           */
84='8'  /*Colorado             */
16='9'  /*Connecticut          */
51='10' /*Delaware             */
53='11' /*District of Columbia */
59='12' /*Florida              */
58='13' /*Georgia              */
95='15' /*Hawaii               */
82='16' /*Idaho                */
33='17' /*Illinois             */
32='18' /*Indiana              */
42='19' /*Iowa                 */
47='20' /*Kansas               */
61='21' /*Kentucky             */
72='22' /*Louisiana            */
11='23' /*Maine                */
52='24' /*Maryland             */
14='25' /*Massachusetts        */
34='26' /*Michigan             */
```

```
41='27' /*Minnesota                 */
64='28' /*Mississippi               */
43='29' /*Missouri                  */
81='30' /*Montana                   */
46='31' /*Nebraska                  */
88='32' /*Nevada                    */
12='33' /*New Hampshire             */
22='34' /*New Jersey                */
85='35' /*New Mexico                */
21='36' /*New York                  */
56='37' /*North Carolina            */
44='38' /*North Dakota              */
31='39' /*Ohio                      */
73='40' /*Oklahoma                  */
92='41' /*Oregon                    */
23='42' /*Pennsylvania              */
15='44' /*Rhode Island              */
57='45' /*South Carolina            */
45='46' /*South Dakota              */
62='47' /*Tennessee                 */
74='48' /*Texas                     */
87='49' /*Utah                      */
13='50' /*Vermont                   */
54='51' /*Virginia                  */
91='53' /*Washington                */
55='54' /*West Virginia             */
35='55' /*Wisconsin                 */
83='56' /*Wyoming                   */
;
run;


%macro recode(dsn,year);


data cps&year;
length wsal _ yn 8.;
        set &dsn;

        * create UNICON recode vars;
        _popstat=p _ stat;
        age=a _ age;
        _year=&year;
        * _ grdhi is only defined for 91 and earlier so don't bother;
        _grdhi=.;
        grdatn=a _ hga;

        %if &year lt 2003 %then %do;
```

```
       _ race=a _ race;
         if _ race>3 then _ race=3;
  %end;
  %if &year ge 2003 %then %do;
         _ race=PRDTRACE;
         if _ race>3 then _ race=3;
  %end;

  %if &year lt 2003 %then %do;
         _ spneth=(1<=A _ REORGN<=7);
  %end;
  %if &year ge 2003 and &year le 2013 %then %do;
         _ spneth=(1<=PRDTHSP<=5);
  %end;
  %if &year ge 2014 %then %do;
         _ spneth=(1<=PRDTHSP<=8);
  %end;

  _ wkslyr=WKSWORK;
  hrslyr=HRSWK;
  _ incwag=wsal _ val;

  if wsal _ yn=1 then _ wklywg=( _ incwag/ _ wkslyr);
  sex=a _ sex;
  %if &year le 2003 %then %do;
         wgt=MARSUPWT;
  %end;
  %if &year ge 2004 %then %do;
         wgt=MARSUPWT;
  %end;

  %if &year le 2000 %then %do;
         state=input(put(hg _ st60,newstate.),2.);
  %end;
  %if &year ge 2001 %then %do;
         state=gestfips;
  %end;

  wgtfnl=A _ FNLWGT;

  keep _ popstat age _ year grdatn _ race _ spneth _ wkslyr hrslyr _ incwag
  sex _ wklywg wgt wgtfnl wsal _ yn state prdthsp prdtrace;

run;
proc contents data=cps&year;
run;
```

```
%mend;

%recode(jum.cpsmar99,1999);
%recode(jum.cpsmar00,2000);
%recode(jum.cpsmar01,2001);
%recode(jum.cpsmar02,2002);
%recode(jum.cpsmar03,2003);
%recode(jum.cpsmar04,2004);
%recode(jum.cpsmar05,2005);
%recode(jum.cpsmar06,2006);
%recode(jum.cpsmar07,2007);
%recode(jum.cpsmar08,2008);
%recode(jum.cpsmar09,2009);
%recode(jum.cpsmar10,2010);
%recode(jum.cpsmar11,2011);
%recode(jum.cpsmar12,2012);
%recode(jum.cpsmar2013,2013);
%recode(jum.cpsmar2014t,2014);
%recode(jum.cpsmar2015,2015);
%recode(jum.cpsmar2016,2016);
%recode(jum.cpsmar2017,2017);
%recode(jum.cpsmar2018,2018);
%recode(jum.cpsmar2019select,2019);

data jum.cps9919;
      set cps1999
            cps2000 cps2001 cps2002 cps2003 cps2004 cps2005 cps2006 cps2007 cps2008
cps2009
            cps2010 cps2011 cps2012 cps2013 cps2014 cps2015 cps2016 cps2017 cps2018
cps2019;
run;
proc contents data=cpssas.cps9919;
run;
```

Combine Recent CPS Data to Previous CPS Data Back to 1980
Inputs: cps8014; cps9919
Output: stackcps8019
Program name: "cpsconsistency_outfile_mar19.sas"

```
libname cpssas "@input _ directory";

proc format;
value any
1-high = '1+';
```

```
value educ
  1="dropout"
  2="hs grad"
  3="some college"
  4="college grad"
  5="college plus";

value race
  1="white"
  2="black"
  3="other"
  4="hispanic";

value sex
  1="Male"
  2="Female";

value UH _ GRDATN _ 1 _ f
   -01 = "Blank"
   031 = "Less than 1st grade"
   032 = "1st, 2nd, 3rd, or 4th grade"
   033 = "5th or 6th grade"
   034 = "7th or 8th grade"
   035 = "9th grade"
   036 = "10th grade"
   037 = "11th grade"
   038 = "12th grade - no diploma"
   039 = "High school graduate - diploma or equivalent (GED)"
   040 = "Some college but no degree"
   041 = "Associate degree - occupational/vocational"
   042 = "Associate degree - academic program"
   043 = "Bachelor's degree (BA, AB, BS, etc.)"
   044 = "Master's degree (MA, MS, MEng, MEd, MSW, etc.)"
   045 = "Professional school degree (MD, DDS, DVM, etc.)"
   046 = "Doctoral degree (PhD, EdD, etc.)"
   -99 = "Missing"
   ;

  value UH _ GRDCOM _ 1 _ f
  01 = "Yes"
  02 = "No"
  -9 = "Missing"
;

value cpi    /* CPI-U for march cps years (actual year in cpi is cps year lagged
since wages are reported for prior year) */
```

```
/*so the value listed here as 2018 is actually for 2017 according to bls, which
works because the 2018 cps data actually covers 2017*/
    1971='38.8'
    1972='40.5'
    1973='41.8'
    1974='44.4'
    1975='49.3'
    1976='53.8'
    1977='56.9'
    1978='60.6'
    1979='65.2'
    1980='72.6'
    1981='82.4'
    1982='90.9'
    1983='96.5'
    1984='99.6'
    1985='103.9'
    1986='107.6'
    1987='109.6'
    1988='113.6'
    1989='118.3'
    1990='124'
    1991='130.7'
    1992='136.3'
    1993='140.3'
    1994='144.5'
    1995='148.2'
    1996='152.4'
    1997='156.9'
    1998='160.5'
    1999='163'
    2000='166.6'
    2001='172.2'
    2002='177.1'
    2003='179.9'
    2004='184'
    2005='188.9'
    2006='195.3'
    2007='201.6'
    2008='207.342'
    2009='215.303'
    2010='214.537'
    2011='218.056'
    2012='224.939'
    2013='229.594'
    2014='232.957'
```

```
   2015='236.736'
   2016='237.017'
   2017='240.007'
   2018='245.12';

VALUE p_stat          (default=32)
        1         =   "Civilian 15+"
        2         =   "Armed Forces"
        3         =   "Children 0 - 14"

run;

data file1;
set cpssas.cps8014;
year=substr(_year,1,4)+0;

wgt_even=wgt/100;
wgtfnl_even=wgtfnl/100;
run;

data file2;
      format wsal_yn prdtrace prdthsp 8.;
set cpssas.cps9919;

year=_year;

wgt_even=wgt;
wgtfnl_even=wgtfnl/1;
run;

****************************************************************;
*** Stack files                       ;
****************************************************************;
data prep1;
set file1;
if year<=2013;
run;
data prep2;
set file2;
if year>=2014;
run;

data stack(drop=prdthsp prdtrace _spneth);
set prep1(keep=year wgt_even wgtfnl_even sex _popstat age grdatn _race _
spneth _wkslyr hrslyr _incwag _wklywg _grdhi grdcom hisp hours _wkstat)
```

```
     prep2(keep=year wgt_even wgtfnl_even sex_popstat age grdatn_race_
spneth_wkslyr hrslyr_incwag_wklywg prdthsp prdtrace);
format sex sex. grdatn uh_grdatn_1_f._race race. grdcom UH_GRDCOM_1_f.
_popstat p_stat.;

label
wgt_even = "CPS Weight"
wgtfnl_even = "CPS Weight Final Available 1976 forward";

*Prior to 1992, the education data lists highest grade attended and whether or
not that grade was completed;
*For 1992 forward, the grade/degree attainted is provided;
*A single longitudinal educational variable called DVD_EDUC is created using
    1980-1991 _grdhi and grdcom
    1992-2018 grdatn;
*Also creating a highest grade attained variable for just 1970-1991 DVD_GRDHI;

format dvd_educ educ.;
if year<=1991 then do;
    if (_grdhi<12 or (_grdhi=12 and grdcom ne 1))     then dvd_educ=1; *
dropout;
    else if (_grdhi=12 and grdcom=1) then dvd_educ=2; * high school;
    else if (13<=_grdhi<16 or (_grdhi=16 and grdcom ne 1)) then dvd_educ=3; *
some college;
    else if (_grdhi=16 and grdcom=1) then dvd_educ=4; * college grad;
    else if (_grdhi>16) then dvd_educ=5; * college plus;
end;
else if year>=1992 then do;
        if (grdatn<=38)     then dvd_educ=1; * dropout;
    else if (grdatn=39)     then dvd_educ=2; * high school;
    else if (40<=grdatn<=42) then dvd_educ=3; * some college;
    else if (grdatn=43)     then dvd_educ=4; * college grad;
    else if (grdatn>43)     then dvd_educ=5; * college plus;
end;
label dvd_educ = "Derived education variable";

if grdcom eq 1 then dvd_grdhi=_grdhi;
else if _grdhi>=1 then dvd_grdhi=_grdhi-1;
label dvd_grdhi = "Derived high grade complete up to 1991";

        if (1980<=year<=1988) and (2<=_spneth<=7) then dvd_hispanicflag=1;
    else if (1989<=year<=2002) and (1<=_spneth<=7) then dvd_hispanicflag=1;
    else if (2003<=year<=2013) and (3<=_spneth<=7) then dvd_hispanicflag=1;
    else if (2004<=year     ) and (1<=prdthsp<=8) then dvd_hispanicflag=1;
    label dvd_hispanicflag = "Derived hispanic flag adjusted for data changes";
```

```
format dvd_newrace race.;
if year>=1980 then do;
      if _race=1 and dvd_hispanicflag=1 then dvd_newrace=4;
      else dvd_newrace=_race;
      label dvd_newrace = "Derived race/ethnicity using _race and
dvd_hispanicflag";
end;


realyear=year-1;

cpi=put(year,cpi.)*1;
wklywgadj=_wklywg*put(2018,cpi.)/cpi;


run;


data cpssas.stackcps8019;
set stack;
run;
proc contents data=cpssas.stackcps8019;
run;
proc export data=cpssas.stackcps8019
            file="/SASdata/Derived/martini/small/wagetabs/stackcps8019.dta"
            dbms=STATA replace;
run;
```

Code to Read in SOFS Data and Output Education Distribution

We read in the SOFS and use linear regressions to smooth the education distribution for each year using Stata. We first do this for enlisted and then for officers.

Read in and Smooth SOFS Enlisted Data

Inputs: "Education Distribution.xlsx" based on the SOFS data; "Greenbook Enlisted Distribution.xlsx" based on Table A6 from the Greenbook

Output: "educ_sofs.dta"

Program name: "SOFS Education Distribution.do"

```
clear all
cap log close

global s = "\"
global rawdata  "@input_directory"
global out "@output_directory"

cap log using "$out${s}SOFS_Ed_distribution_log",t replace //creates a log
file for the program

*Read-in each year separately and create a tempfile
foreach y in 2002 2004 2006 2008 2010 2012 2014 2016 2017 2018{
        import excel "$rawdata${s}Education Distribution.xlsx", clear sheet("`y'")
        firstrow

*Convert the data to numeric
drop if rank=="none"

foreach v of varlist _all{
                replace `v'="0" if `v'=="NR"
        }

        foreach var in dropout hs_grad 11y some_college aa_degree bachelors
        masters_plus{
        destring `var', ignore("%") replace
        replace `var'=`var'/100
        replace `var'=round(`var', 0.01)
        format `var' %9.0g
        }

*Create a tempfile for each year to use later
tempfile year`y'
save `year`y'', replace

*Import Table A6 of the number of enlisted by rank and YOS from Greenbook
```

```
        import excel "$rawdata${s}Greenbook Enlisted Distribution.xlsx", clear
        sheet("`y'") firstrow

drop if rank==""

*Merge to SOFS data
merge 1:1 rank using `year`y''

*Creating a total for each YOS
preserve

collapse (sum) yos*, by(_merge)
reshape long yos, i(_merge) j(j)
rename yos sum
rename j yos
drop _merge

tempfile sum
save `sum', replace

restore

        *Creating a weighted average of number of people in a given YOS weighted
        by the percent in each education category
        foreach ed in dropout hs_grad 11y some_college aa_degree bachelors
        masters_plus{
        preserve

        foreach yos of varlist yos*{
                replace `yos'=`yos'*`ed'
                }

        collapse (sum) yos*, by(_merge)

        reshape long yos, i(_merge) j(j)
        rename yos d
        rename j yos
        drop _merge

        merge 1:1 yos using `sum'
        gen `ed'=d/sum
        keep yos `ed'

        tempfile `ed'
        save "`ed'", replace
```

```
        restore
        }


*Merging together the number of people in each education category for each YOS
use "dropout", clear
foreach var in hs_grad 11y some_college aa_degree bachelors masters_plus{
        merge 1:1 yos using "`var'"
        drop _merge
        }


*Combining "less than 1 year of college" and "some college"
gen sc=11y+some_college


*Keep if yos is 30 or less
keep if yos<32


*Creating powers of YOS to use in regressions
gen yos2 = yos*yos
gen yos3 = yos2*yos
gen yos4 = yos3*yos
gen yos5 = yos4*yos
gen yos6 = yos5*yos


summarize


        /* Regressions and predictions by ed level using different degree
        polynomials for each category */
regress hs_grad yos yos2 yos3 yos4 yos5 yos6
predict phs if e(sample)


regress sc yos yos2 yos3 yos4 yos5 yos6
predict psc if e(sample)


regress aa_degree yos yos2 yos3 yos4 yos5
predict paa if e(sample)


regress bachelors yos yos2 yos3 yos4
predict pba if e(sample)


regress masters_plus yos yos2 yos3 yos4
predict pma if e(sample)


*Combine sc and aa
gen sc_aa=psc+paa


replace pma=0 if pma<0
```

```
*Normed to sum to 1.00
gen sum=phs+sc_aa+pba+pma
replace phs=phs/sum
replace sc_aa=sc_aa/sum
replace pba=pba/sum
replace pma=pma/sum

keep yos phs sc_aa pba pma
rename phs ed2
rename sc_aa ed3
rename pba ed4
rename pma ed5

*Changing format and saving to be combined together and merged in next program
reshape long ed, i(yos) j(j)
rename ed ed_percent
rename j dvd_educ
gen year=`y'
tempfile educ`y'
save `educ`y''
}

*Combining all years to one file
use `educ2002', clear
foreach year in 2004 2006 2008 2010 2012 2014 2016 2017 2018{
append using `educ`year''
}

save "$rawdata${s}educ_sofs", replace
cap log close
```

Read in and Smooth SOFS Officer Data

Inputs: "Education Distribution - Officers.xlsx" based on the SOFS data; "Greenbook Officer Distribution.xlsx" based on Table A6 from the Greenbook

Output: "educ_sofs.dta"

Program name: "SOFS Education Distribution - Officers.do"

```
clear all
cap log close
global s = "\"
global rawdata  "@input_directory"
global rawdata2  "@input_directory"
global out "@output_directory "
cap log using "$out${s}SOFS_Ed_distribution_officers_log",t replace //creates
a log file for the program
```

```
foreach y in 2002 2004 2006 2008 2010 2012 2014 2016 2017 2018{

        import excel "$rawdata${s}Education Distribution - Officers.xlsx", clear
        sheet("`y'") firstrow

drop if rank=="none" | rank==""

foreach v of varlist _all{
            replace `v'="0" if `v'=="NR"
        }

foreach var in hs_grad 11y some_college coll_grad masters_plus{
        destring `var', ignore("%") replace
        replace `var'=`var'/100
        replace `var'=round(`var', 0.01)
        format `var' %9.0g
        }

tempfile year`y'
save `year`y'', replace

        import excel "$rawdata${s}Greenbook Officer Distribution.xlsx", clear
        sheet("`y'") firstrow

drop if rank==""
merge 1:1 rank using `year`y''
keep if _merge==3

preserve

collapse (sum) yos*, by(_merge)

reshape long yos, i(_merge) j(j)
rename yos sum
rename j yos
drop _merge

tempfile sum
save `sum', replace

restore

foreach ed in hs_grad 11y some_college coll_grad masters_plus{
            preserve
        foreach yos of varlist yos*{
```

```
                    replace `yos'=`yos'*`ed'
           }

       collapse (sum) yos*, by(_merge)
       reshape long yos, i(_merge) j(j)
       rename yos d
       rename j yos
       drop _merge

       merge 1:1 yos using `sum'
       gen `ed'=d/sum
       keep yos `ed'

       tempfile `ed'
       save "`ed'", replace

       restore
       }

use "hs_grad", clear
foreach var in 11y some_college coll_grad masters_plus{
       merge 1:1 yos using "`var'"
       drop _merge
       }

keep if yos<32

gen yos2 = yos*yos
gen yos3 = yos2*yos
summarize

/* Edpct regressions and predictions by ed level */
regress masters_plus yos yos2 yos3
predict pma if e(sample)

*Normed to sum to 1.00
gen pba=1-pma

keep yos pba pma
rename pba ed4
rename pma ed5

reshape long ed, i(yos) j(j)
rename ed ed_percent
rename j dvd_educ
gen year=`y'
```

```
tempfile educ`y'
save `educ`y''
}

use `educ2002', clear
foreach year in 2004 2006 2008 2010 2012 2014 2016 2017 2018{
append using `educ`year''
}

save "$rawdata${s}educ_sofs_officer", replace
cap log close
```

Code to Create Weighted Average RMC over Time

We create RMC percentiles for each category of civilian education. We then weight each RMC percentile according to the number of enlisted and officers in that education and YOS category. Finally, we constructed a weighted average over YOS 1–20.

Constructing a Weighted Average over Time for Enlisted Personnel

Inputs: "educ_sofs.dta"; "cpi.xlsx" from BLS; "Gender Mix.xlsx" based on Pop Rep; "Greenbook Numbers by year, 1975 - 2019.xlsx" based on Greenbook Table A6; "stackcps8019.dta"; "Greenbook RMC by year, 1973 - 2019.xlsx" based on Greenbook Table B4
Outputs: "weighted_greenbooks.xlsx" - RMC percentiles by education and YOS in Excel; "Enlisted RMC by Year_sofs.png" – Graph of overall weighted RMC percentiles by year
Program name: "Weighted Average over Time.do"

```
clear all
cap log close
global s = "\"
global rawdata  "@input_directory"

global out "@output_directory"

cap log using "$out${s}Weighted_Average_log",t replace //creates a log file for
the program

*Creating a CPI file to adjust for inflation. CPI-U Data from: https://data.bls.
gov/timeseries/cuur0000sa0?series_id=cwur0000s
import excel "$rawdata${s}cpi.xlsx", clear firstrow
rename Year year
rename Annual cpi
save "$rawdata${s}cpi", replace

*Importing data on gender mix from Appendix D of Pop Rep
```

```
import excel "$rawdata${s}Gender Mix.xlsx", clear firstrow
gen fp_enl=enl_female_per/100
gen fp_off=off_female_per/100
gen mp_enl=1-fp_enl
gen mp_off=1-fp_off
drop enl_female_per off_female_per
rename enl_female tot_fem_enlisted
rename off_female tot_fem_officer
save "$rawdata${s}gender_mix", replace

*Importing Greenbook Data
import excel "$rawdata${s}Greenbook Numbers by year, 1975 - 2019.xlsx",
sheet("ALLENL") clear firstrow
rename Under_2 yos_1
rename *, lower
drop total z

*YOS 1 - 4 and then jumps to every other year
forvalues i = 1/4 {
rename yos_`i' yos`i'
}
forvalues i = 6(2)40 {
rename yos_`i' yos`i'
}

reshape long yos, i(year pay) j(j)
rename pay rank
rename yos number
rename j yos
save "$rawdata${s}total_enlisted", replace

*Importing CPS data
use "$rawdata${s}stackcps8019.dta", clear

*Relabeling the highest education category
label define dvd_educ 5 "masters plus", modify

*Aligning CPS year to year of data from Greenbooks
rename year cps_year
rename realyear year
rename wklywgadj wklywgadj2017

*Data was originally in 2017 dollars; we will convert to 2019 dollars
drop cpi

merge m:1 year using "$rawdata/cpi"
```

```
keep if _merge==3
drop _merge

*Inflating pay to 2019 dollars
gen wklywgadj=_wklywg*Inflationfactor

*Drop earlier years
keep if year>1974

*Merge in Military Gender Mix
merge m:1 year using "$rawdata${s}gender_mix"
drop _merge

*Subset on civilian adults (aged 15+; nonmilitary)
keep if _popstat==1

*A full-time, full-year worker is one with a usual work week of more than 35
hours and who worked more than 35 weeks in the year
drop if hrslyr<36
drop if _wkslyr<36

*Generating an estimated YOS (in this case years of experience) based on age
and education level
gen yos = age - 18 if dvd_educ==1
replace yos = age - 18 if dvd_educ==2
replace yos = age - 20 if dvd_educ==3
*replace yos = age - 20 if dvd_educ=="aa degree"
replace yos = age -22 if dvd_educ==4
replace yos = age - 24 if dvd_educ==5
drop if yos<0

*Weighting according to military gender mix
gen wgt_o=wgt_even*mp_off if sex==1
replace wgt_o=wgt_even*fp_off if sex==2

gen wgt_e=wgt_even*mp_enl if sex==1
replace wgt_e=wgt_even*fp_enl if sex==2
drop if wklywgadj==.

*Replacing yos to match the Greenbook yos which only have every other year
after year 4.
replace yos=1 if yos==0
replace yos=4 if yos==5
replace yos=6 if yos==7
replace yos=8 if yos==9
replace yos=10 if yos==11
```

```
replace yos=12 if yos==13
replace yos=14 if yos==15
replace yos=16 if yos==17
replace yos=18 if yos==19
replace yos=20 if yos==21
replace yos=22 if yos==23
replace yos=24 if yos==25
replace yos=26 if yos==27
replace yos=28 if yos==29
replace yos=30 if yos==31

save "$rawdata${s}cps", replace

*Importing Greenbook Data
import excel "$rawdata${s}Greenbook RMC by year, 1973 - 2019.xlsx",
sheet("ALLENL") clear firstrow

*Restructuring data to be able to more easily make use of it
rename Under _ 2 yos _ 1
rename *, lower

forvalues i = 1/4 {
rename yos _ `i' yos`i'
}

forvalues i = 6(2)40 {
rename yos _ `i' yos`i'
}

reshape long yos, i(year pay note) j(j)
rename pay rank
rename yos pay
rename j yos

*Merging in CPI data
merge m:1 year using "$rawdata/cpi"
keep if _ merge==3
drop _ merge

*Inflating pay to 2019 dollars
gen pay _ adj=(pay*Inflationfactor)/52

keep year yos pay _ adj

*Merge in CPS data
merge 1:m year yos using "$rawdata${s}cps"
```

```
keep if _merge==3
drop _merge

*Defining the group of interest for RMC comparison - education, YOS, and year
egen category = group(dvd_educ yos year)

sort category

*Creating percentile of civilian pay for the given RMC by group
*Note that the relrank command does not come standard in Stata and must be
downloaded. *If your version of Stata has access to the Internet, this can be
installed by typing the following into the command window: ssc install relrank
by category:relrank pay_adj, ref(wklywgadj [aw=wgt_e]) g(rmc_p)

gen count=1

*Getting the 50th and 70th percentiles of the civilian wage distribution
collapse (p50) p50=wklywgadj (p70) p70=wklywgadj (sum) count [w=wgt_e], by(year
yos dvd_educ pay_adj rmc_p)

gen rmc_per=round(rmc_p*100)

*Labeling variables
label variable p50 "P_50"
label variable p70 "P_70"
label variable pay_adj "YOS RMC from Geenbooks"
label variable yos "Years of Service/Years of Experience"

*Exporting data to Excel
export excel "$out/weighted_greenbooks.xlsx", replace firstrow(variables)

decode dvd_educ, gen(education)
gen education_label=proper(education)

*Merge in education data
*Using Status of Forces data for years available
merge m:1 year yos dvd_educ using "$rawdata${s}educ_sofs"
keep if _merge==3

*Creating a weighted average
collapse (mean) rmc_p [weight=ed_percent], by(year yos)

merge 1:1 year yos using "$rawdata${s}total_enlisted"
keep if _merge==3
drop if yos>20
```

```
collapse (mean) rmc _ p [weight=number], by(year)

gen rmc=rmc _ p*100
gen rmc _ label=round(rmc)


*Graphing final percentiles
twoway line rmc year, lcolor(black) || scatter rmc year, mlabel(rmc _ label)
mlabposition(12) mcolor(black) msize(vsmall) mlabcolor(black) mlabsize(vsmall)
ylabel(40(10)100, labsize(vsmall) angle(0)) xlabel(1980(2)2018, labsize(vsmall))
leg(off) ytitle("Enlisted Weighted Average RMC") title("Enlisted Weighted Average
RMC by Year, Payfiles (dropout)") graphregion(color(white)) bgcolor(white)

graph export "$out/Enlisted RMC by Year _ sofs.png", width(1000) replace
cap log close
```

Constructing a Weighted Average over Time for Officers

Inputs: "educ_sofs_officer.dta"; "cpi.xlsx" from BLS; "Gender Mix.xlsx" based on Pop Rep; "Greenbook Numbers by year, 1975 - 2019.xlsx" based on Greenbook Table A6; "stackcps8019. dta"; "Greenbook RMC by year, 1973 - 2019.xlsx" based on Greenbook Table B4
Outputs: "weighted_greenbooks_officer.xlsx" - RMC percentiles by education and YOS in Excel; "Officer RMC by Year_sofs.png" – Graph of overall weighted RMC percentiles by year
Program name: "Weighted Average over Time - Officers.do"

```
clear all
cap log close
global s = "\"
global rawdata "@input _ directory"

global out "@output _ directory"

cap log using "$out${s}Weighted _ Average _ Officers _ log",t replace //creates a
log file for the program

*Importing Greenbook Data on total numbers for officers
import excel "$rawdata${s}Greenbook Numbers by year, 1975 - 2019.xlsx",
sheet("ALLCO") clear firstrow
rename Under _ 2 yos _ 1
rename *, lower
drop total z

forvalues i = 1/4 {
rename yos _ `i' yos`i'
}

forvalues i = 6(2)40 {
rename yos _ `i' yos`i'
```

```
}

reshape long yos, i(year pay) j(j)
rename pay rank
rename yos number
rename j yos

save "$rawdata${s}total_officer", replace

*Importing Greenbook Data of pay for officers
import excel "$rawdata${s}Greenbook RMC by year, 1973 - 2019.xlsx", sheet("ALLCO")
clear firstrow

rename Under_2 yos_1
rename *, lower

forvalues i = 1/4 {
rename yos_`i' yos`i'
}
forvalues i = 6(2)40 {
rename yos_`i' yos`i'
}

reshape long yos, i(year pay note) j(j)
rename pay rank
rename yos pay
rename j yos

merge m:1 year using "$rawdata/cpi"
keep if _merge==3
drop _merge

*Inflating pay to 2019 dollars
gen pay_adj=(pay*Inflationfactor)/52

keep year yos pay_adj

*Merge in CPS data
merge 1:m year yos using "$rawdata${s}cps"
keep if _merge==3
drop _merge

egen category = group(dvd_educ yos year)
sort category
```

```
*Note that the relrank command does not come standard in Stata and must be
downloaded. *If your version of Stata has access to the Internet, this can be
installed by typing the following into the command window: ssc install relrank
by category:relrank pay_adj, ref(wklywgadj [aw=wgt_o]) g(rmc_p)

gen count=1

collapse (p50) p50=wklywgadj (p70) p70=wklywgadj (sum) count [w=wgt_o], by(year
yos dvd_educ pay_adj rmc_p)

gen rmc_per=round(rmc_p*100)

label variable p50 "P_50"
label variable p70 "P_70"
label variable pay_adj "YOS RMC from Geenbooks"
label variable yos "Years of Service/Years of Experience"

export excel "$out/weighted_greenbooks_officer.xlsx", replace
firstrow(variables)

decode dvd_educ, gen(education)
gen education_label=proper(education)

*Merge in education data
*Using Status of Forces data for years available
merge m:1 year yos dvd_educ using "$rawdata${s}educ_sofs_officer"
keep if _merge==3

collapse (mean) rmc_p [weight=ed_percent], by(year yos)

merge 1:1 year yos using "$rawdata${s}total_officer"
keep if _merge==3
drop if yos>20

collapse (mean) rmc_p [weight=number], by(year)

gen rmc=rmc_p*100
gen rmc_label=round(rmc)

twoway line rmc year, lcolor(black) || scatter rmc year,  mlabel(rmc_label)
mlabposition(12) mcolor(black) msize(vsmall) mlabcolor(black) mlabsize(vsmall)
ylabel(40(10)100, labsize(vsmall) angle(0)) xlabel(1980(2)2018, labsize(vsmall))
leg(off) ytitle("Officer Weighted Average RMC") title("Officer Weighted Average
RMC by Year, SOFS") graphregion(color(white)) bgcolor(white)

graph export "$out/Officer RMC by Year_sofs.png", width(1000) replace
cap log close
```

Additional Defense Employment Cost Index Tabulations

Comparison of the CPS ORG Versus the ASEC Earnings Data

Figure D.1 illustrates the relative timings of ORG versus the ASEC earnings data with respect to a given October, the month when the current process of setting the annual pay raise for the following fiscal year begins with the release of the Fall ECI data. The leftmost column depicts the timing of the ASEC subsample, who are asked in March to recall earnings over the prior 12 months. The six columns on the right depict the timing of the earnings measures collected from the ORG subsample between April and September. The earnings data derived from the ORG subsample are between one and six months old when the review of the data begins, while the ASEC earnings data are between seven and 18 months old with respect to this reference month. Finally, in the series of validation exercises we conducted, we found that the DECI using the ORG data consistently performed at least modestly better than the DECI using ASEC data.

Empirically, the choice of CPS subsample is not trivial. Figure D.2 follows the path of DECIs constructed using both ASEC and ORG data. On average, a DECI constructed using weekly earnings derived from annual earnings estimates in the ASEC data suggests modestly higher wage growth. This relationship holds across approximately the entire sample period, though the magnitude of the average gap between the two indices has grown substantially since the early 2000s. The largest single-year differences in measured earnings changes are around 0.3 percentage points (in 2016, for example, the year-to-year earnings change was 2.2 percent using the ASEC data and 1.9 percent using the ORG data). Over the nearly 40-year sample period, the difference in the final index values amounts to around 10 percent of the DECI index value.

A likely candidate for the source of this disagreement in earnings growth is the over-reporting of annual income relative to reported weekly earnings. In a comparison of reported annual wage and salary income among respondents appearing in the ASEC subsample who also were in the ORG subsample in March, we find that imputing annual earnings using the product of a respondent's weekly earnings reported as part of the ORG earnings supplement data and weeks worked in the prior year from the ASEC data are around 2 to 5 percent higher on average across the sample period. This persistent difference may be due to recall error on the part of respondents asked to estimate annual wage and salary earnings or other factors related to aggregating work history over time.

Figure D.1
Timing Differences Between CPS Subsamples

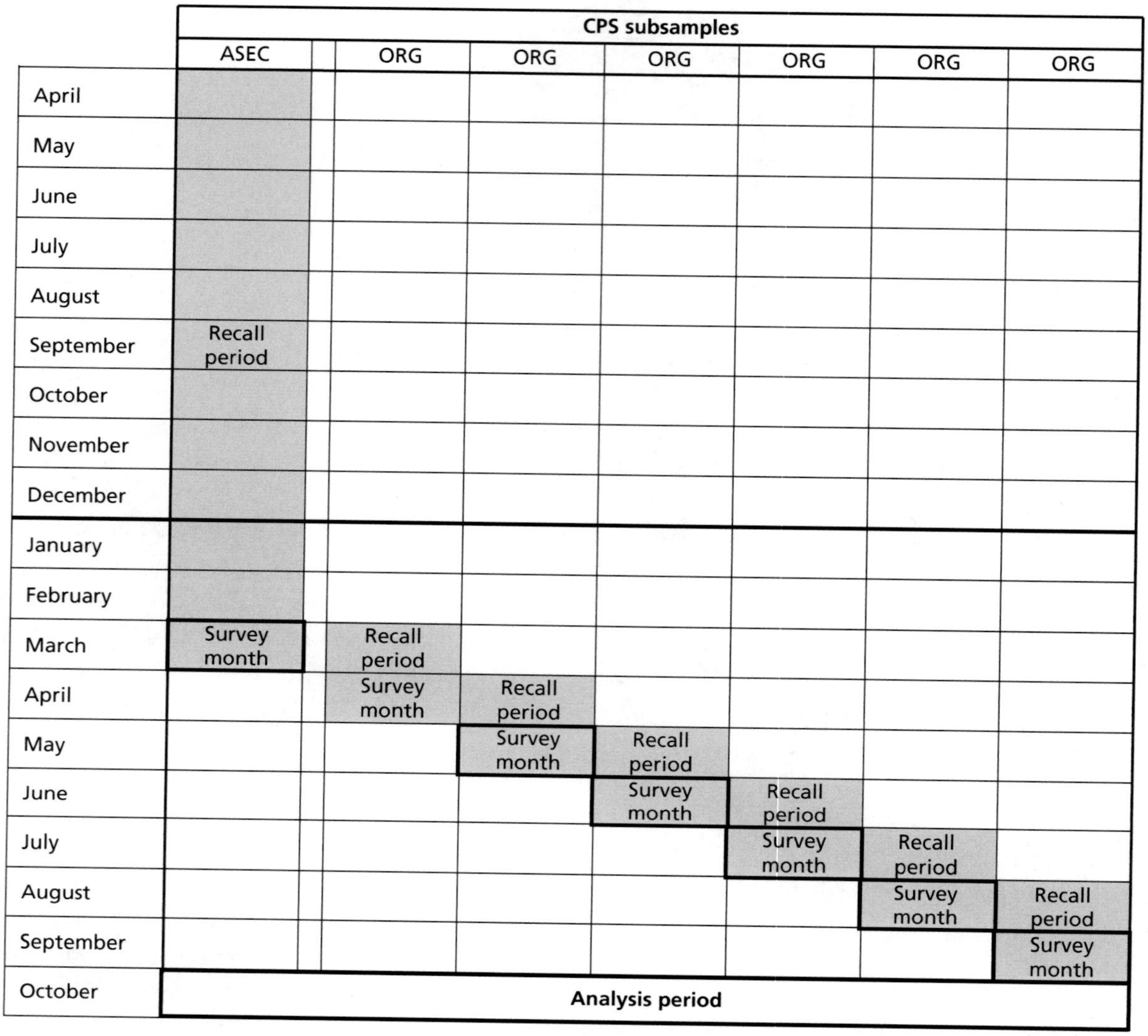

	CPS subsamples						
	ASEC	ORG	ORG	ORG	ORG	ORG	ORG
April							
May							
June							
July							
August							
September	Recall period						
October							
November							
December							
January							
February							
March	Survey month	Recall period					
April		Survey month	Recall period				
May			Survey month	Recall period			
June				Survey month	Recall period		
July					Survey month	Recall period	
August						Survey month	Recall period
September							Survey month
October	Analysis period						

Comparing Our DECI Results with Hosek 1992

To ensure that we are following a suitably similar approach to constructing the DECI as the original study and to assess the effects of the alternative modeling choices discussed above, we examined how well we were able to approximate results from Hosek 1992, which focused on the years 1977 to 1991. Because of data limitations, we focused on the period from 1982 to 1991. As with the original study, we used 1982 as the baseline year for computing the indices.

The Basic Pay Index: A Tool to Measure Military/Civilian Pay Gaps

In Hosek 1992, a key approach used to illustrate the implications of the ECI or the DECI for military pay growth was to plot a standardized measure of the gap (or surplus) in the difference between an index of military pay, the "basic pay index" (BPI), and either the ECI or the DECI. The BPI is generated by starting with a value of 100 in the chosen baseline year and

Figure D.2
Comparing DECIs Constructed Using Different CPS Subsamples

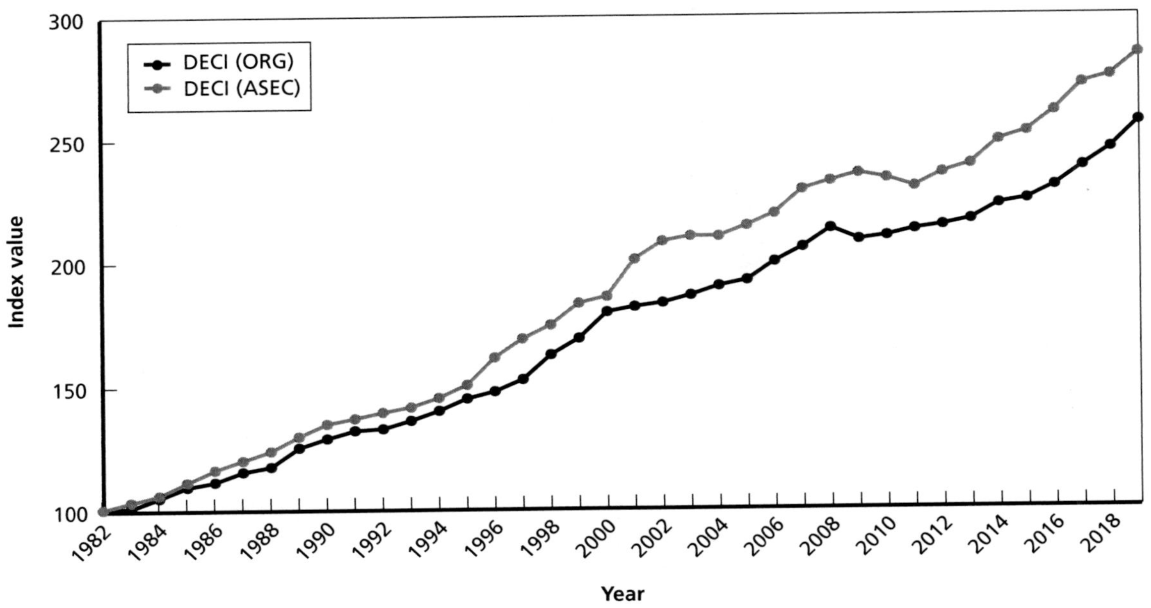

SOURCES: CPS ORG data (April–September, 1982–2019) and CPS ASEC data (March 1982–2019) from IPUMS.
NOTE: Military compositional weights use ADMF data from DMDC. Baseline year for these indices is 1982, which is set equal to 100.

then multiplying this value by the percentage change in basic pay in each subsequent year.[1] The gap measure constructed in Hosek 1992 was the following ratio (using the ECI as the civilian index): (BPI – ECI)/BPI. This ratio measures the distance between the BPI and the comparison index (ECI or DECI) as a *proportion* of the level of the BPI (i.e., percentage divided by 100). To fix ideas, Figure D.3 presents a side-by-side comparison of our current estimated earnings gaps using the ECI index and those reported in Hosek 1992.

Here we used virtually the same data (allowing for small differences in how we calculate basic pay increases when there were differential increases for specific pay grades and also for small revisions over time to the ECI series), so our results were virtually identical. Since both indices were set to 100 in 1982, the estimated gap was zero in that year, and this figure tracked the relative annual pay gap implied by the ECI from the following year, 1983, until 1991. Using the ECI implied that a substantial pay gap arose quickly after 1982 and grew from around 6 to 8 percent in the mid-1980s to 10 to 12 percent by the early 90s. The magnitude of this ECI-based estimated pay gap during a time when recruiting and retention goals were regularly met was a primary piece of the puzzle motivating the exploration of an alternate index to use in adjusting military pay.

In Figure D.4, we present the original estimates of the DECI-based pay gap from Hosek 1992 and our results over this same period. It is important to note that for this exercise we used the ASEC subsample of CPS respondents, as was done in Hosek 1992, but we omitted

---

[1]   Thus, for a 3 percent increase in basic pay in the year after the chosen baseline year, the BPI would be 100.00*1.03 = 103.00. For a 5 percent increase in the following year, the BPI would increase to 103.00*1.05 = 108.15 and so on.

Figure D.3
Comparing ECI-Based Earnings Differences

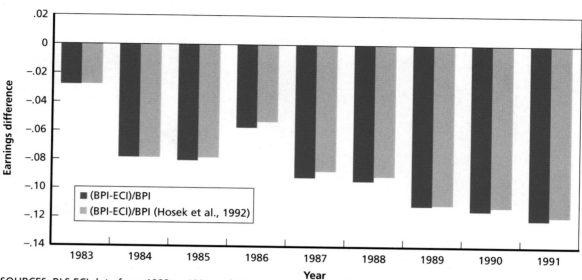

SOURCES: BLS ECI data from 1982 to 1991 and Hosek et al. (1992). BPI data from OUSD(P&R) (2018).
NOTE: The formula for the earnings gap is given in the legend above and it is measured as a proportion (i.e., percent/100), so that 0.2 equals 20 percent.

Figure D.4
Comparing DECI-Based Earnings Differences with Hosek 1992 Using the CPS ASEC Subsample

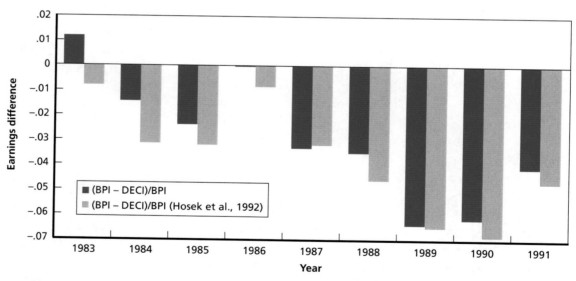

SOURCES: CPS data (IPUMS), BLS, OUSD(P&R) (2018), and Hosek et al. (1992).
NOTES: DECI cells are stratified by eight age groups, four education groups, and gender. Hosek 1992 used three education groups and additionally used six occupational groups. We omit occupation. Military weights generated from ADMF data by aggregating officers and enlisted, omitting Navy officers because of poor coding of education. BPI data from OUSD(P&R) (2018). The formula for the earnings gap is given in the legend above and is measured as a proportion (i.e., percent/100), so that 0.2 equals 20 percent.

occupation, which they included, used a fourth education category, and weighted by gender, as discussed earlier. Despite these differences, our results were remarkably similar to the original DECI-based pay gap measured in Hosek 1992, though the pay gap we estimated was slightly smaller on average. More notable than the differences between our estimates and those from Hosek 1992 though, is the fact that the overall pay gap implied by either approach to constructing the DECI was approximately half the magnitude of the pay gap implied by the ECI, suggesting that the DECI measures a much different path of earnings growth. In Hosek 1992, the authors conjectured that multiple factors related to the data difference discussed in Chapter Four—including the negative shock to wages represented by the large influx of Baby Boomers into the labor market in the late 1970s and the earliest effects of the globalization of manufacturing jobs—led to lower wages for the age and experience level of workers that are more heavily weighted in the DECI.

In our preferred approach, using the April through September ORG subsample of CPS respondents, we estimated an even smaller pay gap during this period. This comparison is presented in Figure D.5. There are fewer similarities between the broad pattern of more and less negative single-year gaps in Hosek 1992 and the alternating earnings gaps and surpluses in our analysis using the ORG-based DECI. The average gap across all nine years in our analysis was zero, while it was around 3.7 percentage points in the original Hosek 1992 DECI results.

Thus, to the extent that the size of the military pay gap implied by the DECI calculated in Hosek 1992 suggested a fairly substantial pay gap that was difficult to square with the successful accession and retention outcomes over those years, our analysis using a different subsample of the CPS suggested an even smaller pay gap, one that provides less evidence of

Figure D.5
Comparing DECI-Based Earnings Differences with Hosek 1992 Using the CPS ORG Subsample

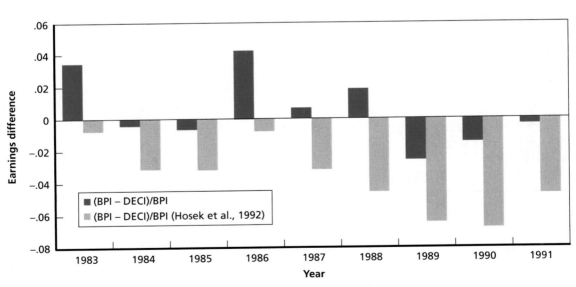

SOURCES: CPS data (IPUMS), BLS, OUSD(P&R) (2018), and Hosek et al. (1992).
NOTES: DECI cells are stratified by eight age groups, four education groups, and gender. Hosek 1992 used three education groups and additionally used six occupational groups. We omit occupation. Military weights generated from ADMF data by aggregating officers and enlisted, omitting Navy officers because of poor coding of education. BPI data from OUSD(P&R) (2018). The formula for the earnings gap is given in the legend above and is measured as a proportion (i.e., percent/100), so that 0.2 equals 20 percent.

a paradox around military pay and manpower outcomes than either the ECI or the original DECI study results.

Additional Sensitivity Analysis Regarding the Choice of Base Year

To further demonstrate the sensitivity of the guidance provided by both of these indices to the chosen baseline year, Table D.1 calculates cumulative percentage changes in the BPI, DECI, and ECI using four different baseline years: 1982, 1990, 2000, and 2010. In Chapter Four, we discussed reasons that 1982 and 2010 might be reasonable baseline years to use in assessing differences in earnings growth. 1990 and 2000 are arbitrary from a policy perspective and were chosen simply because they approximately equally partition our overall sample period. Consistent with the differences shown in Figure 4.6, in 2019 the ECI is nine index points higher than the BPI when using 1982 as a baseline year, but the DECI is 52 index points lower. Using 1990 as the baseline year, the change in *both* civilian earnings indices are lower than the change in the BPI (19 index points lower for the ECI and 45 index points lower for the DECI). A baseline year of 2000 suggests a military earnings surplus according to both civilian earnings indices of similar magnitude (16 index points for the ECI and 35 index points for the DECI). Finally, the implied earnings surplus becomes a small, nearly identical gap of four to five index points when using either index with 2010 as the baseline year.

Using the DECI to Analyze Earnings Growth Among Subgroups

In this subsection, we present comparisons of subgroups DECIs defined by age and educational attainment, as well as enlisted personnel and officers. In these subgroup DECIs, we omit gender weights to avoid generating excessive numbers of empty data cells. Figure D.6 graphically presents earnings differences for subgroup DECIs of educational attainment from the reference point of the two baseline years we have considered above, 1982 and 2010. Panel A, which used 1982 as the baseline year, suggested that there were often substantial earnings gaps for service members with a Baccalaureate degree that persisted until the long period of above-ECI pay increases in the first decade of the 2000s closed them. Around four years later, during the aftermath of the Great Recession, an earnings surplus emerged, potentially

Table D.1
Differences in the BPI, DECI, and ECI over Time, by Baseline Year

To Year	Base Year 1982			Base Year 1990			Base year 2000			Base year 2010		
	BPI	DECI	ECI	BPI	DECI	ECI	BPI	DECI	ECI	BPI	DECI	ECI
1990	26	28	41									
2000	73	79	96	37	39	39						
2010	163	109	162	108	63	85	51	16	33			
2020	208	156	217	144	99	125	77	42	61	17	22	21

NOTES: Basic pay data from OUSD(P&R) (2018). ECI data from the BLS (as described in text). DECI data use CPS ORG data, April–September from 1982 to 2019 and DMDC ADMF data (as described in text).

Figure D.6
DECI-Based Earnings Differences by Education

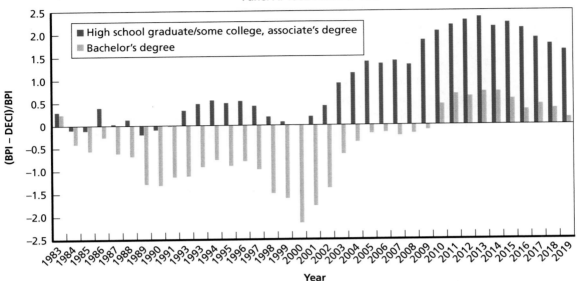

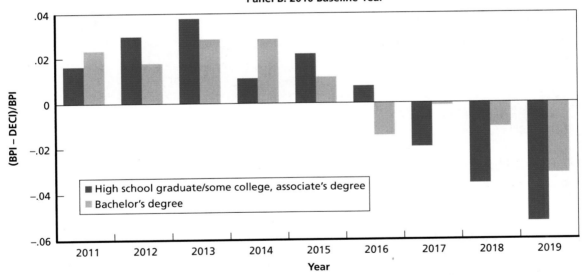

SOURCES: CPS ORG data (April–September 1982–2019) from IPUMS, BPI data from USD(P&R) (2018).
NOTES: DECI cells stratified by eight age groups and four education groups. Military weights are generated from ADMF data by aggregating officers and enlisted, omitting Navy officers because of poor coding of education. The formula for the earnings gap is given on the y-axis above and is measured as a proportion (i.e., percent/100), so that 0.2 equals 20 percent.

related to "upskilling," or employers requiring higher measures of worker skill for a given job in response to a greater pool of potential employees (Modestino, Shoag, and Ballance 2019).

In contrast, for those with a high school diploma, some college, or an associate's degree, the subgroup DECI indicates no systematic gap or surplus until the mid-1990s, when a surplus as large as 6 to 7 percent emerged, then declined and disappeared during the years of the dot-com boom, before reemerging and growing as large as 20 to 25 percent around the early 2010s. Using 2010 as a baseline year, the story is quite different, with these subgroups DECIs suggesting a modest earnings surplus for both groups in the early years of the 2010s that turns into a growing earnings gap for the group with lower educational attainment after 2016.

It is notable that the differences between these two subgroup DECIs using either 1982 or 2010 as the baseline year are qualitatively similar to the differences between the ECI (approximated by the Baccalaureate degree subgroup DECI) and the overall DECI (approximated by the high school/some college/associate's degree subgroup DECI) presented above. In the main text in Chapter Four, we assess the extent to which it is the weighting of the CPS data by the educational attainment composition of military personnel that drives the difference between the ECI and the DECI and find that this is a substantial factor in the difference between the indices.

In Figure D.7, which estimated subgroup DECIs for 17- to 21-year-olds and 32- to 36-year-olds, we see evidence of a modest earnings gap for older individuals through the mid-1990s and earnings gaps for both age groups during the late 1990s. But after this time, the large pay increases of the early 2000s suggest that military pay for both of these age groups grew much more rapidly than the earnings of their civilian counterparts. From the perspective of a 2010 baseline year, these age subgroup DECIs suggest the emergence of an earnings gap in 2014 that grew as large as 10 percent by 2018.

Finally, we compare subgroup DECIs for enlisted personnel and commissioned officers in Figure D.8. Since there is a strong association between enlisted/officer status and educational attainment, it is unsurprising that these results strongly mirror the educational attainment subgroup DECIs in Figure D.6, suggesting that enlisted service members have mostly experienced increases in the rate of pay growth as large earnings surpluses relative to their opportunities in the civilian labor market, whereas officers experienced a persistent gap in earnings growth that was only successfully addressed by the sustained pay increases of the early 2000s. The results using 2010 as the baseline year similarly suggest the recent emergence of a gap in earnings growth between enlisted service members and their civilian counterparts. The strongly negative association between these measures (i.e., "appropriate" growth of enlisted pay suggests earnings gaps for officers and correcting pay gaps for officers suggests significant surpluses for enlisted personnel) suggests that the DECI may provide value not only in informing the annual pay raise, but also in assessing the appropriate structuring of earnings growth within the pay table.

Comparing DECIs Constructed with ADMF and SOFS Data

In Chapter Four, we discuss the issue of measurement error in educational attainment in the ADMF data. We found that most of this error appeared to be related to measured education being smaller than actual education, with this bias being largest with respect to coding service members with some college or an associate's degree as having a high school diploma only. This

Figure D.7
DECI-Based Earnings Differences by Age

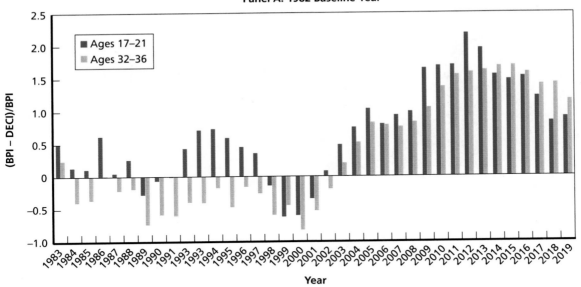

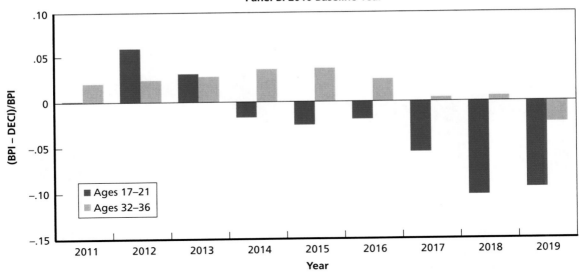

SOURCES: CPS ORG data (April–September 1982–2019) from IPUMS, BPI data from USD(P&R) (2018).
NOTES: DECI cells stratified by eight age groups and four education groups. Military weights are generated from ADMF data by aggregating officers and enlisted, omitting Navy officers because of poor coding of education. The formula for the earnings gap is given on the y-axis above and is measured as a proportion (i.e., percent/100), so that 0.2 equals 20 percent.

Figure D.8
DECI-Based Earnings Differences by Enlisted or Officer Status

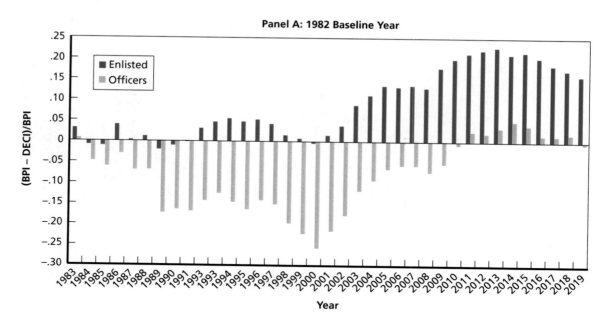

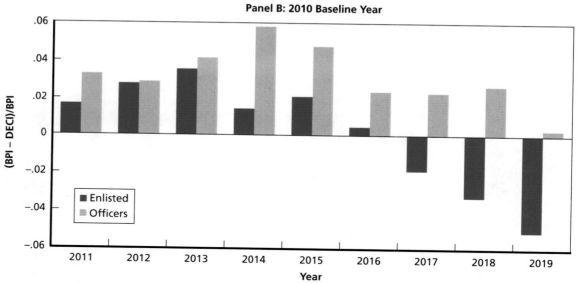

SOURCES: CPS ORG data (April–September 1982–2019) from IPUMS, BPI data from USD(P&R) (2018).
NOTES: DECI cells stratified by eight age groups and four education groups. Military weights are generated from ADMF data separately for enlisted members, and for officers omitting Navy officers because of poor coding of education. The formula for the earnings gap is given on the y-axis above and is measured as a proportion (i.e., percent/100), so that 0.2 equals 20 percent.

fact, along with the evidence that the earnings growth trajectories of these education levels has trended very similarly in the CPS over time, suggests to us that our primary approach of aggregating together service members with a high school diploma, some college, or an associate's degree into a single education grouping largely addresses these issues of measurement error in the data as they pertain to the accuracy of the DECI. However, we also empirically assessed the extent to which this sort of measurement error could substantively affect the guidance provided by the DECI on earnings growth and found any differences related to miscoding of educational attainment to likely be small in magnitude in terms of overall effects on the DECI. In Figure D.9, we present a graphical comparison of the time series of DECIs constructed using, first, the ADMF data with four educational categories and a DECI constructed using tabulations from the biannual SOFS data over the period 2002 to 2018. We found little difference in the path of these DECIs over time and, after the full 16 years in this sample period, the indices differ by only around 3 index points. We believe the advantage of the ADMF data—namely, that is represents a census of military personnel—outweigh the benefits of using a data source like the SOFS, which is a voluntary survey that is collected only biannually and has a sample size that results in both large margins of error regarding group size and significant numbers of cells with zero counts of service members that are represented in the ADMF data.

Figure D.9
Comparison of DECI Weights Generated Using ADMF and SOFS Data

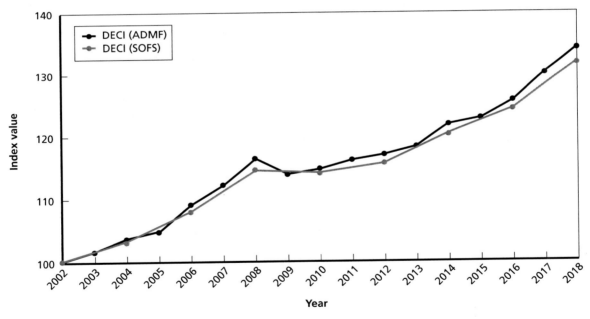

SOURCE: CPS ORG data (April–September 1982–2019) from IPUMS.
NOTES: Military compositional weights from two sources. The first is ADMF data from DMDC. The second is SOFS data from 2002 to 2018. Baseline year for these indices is 2002, which is set equal to 100.

Process for Estimating the Defense Employment Cost Index

In this appendix, we describe the data requirements and the computer coding required to generate the DECI (and we also include a brief description of the other data used in Chapter Four). The description and software code examples reference the requirements to estimate the 2019 DECI using data from 2018 and 2019. To generate updated DECI data for subsequent years, one need only substitute the relevant years (e.g., 2019, 2020, etc.) or add another year to the existing data (e.g., download 2020 and use existing data from 2019) and enter the prior year DECI value, as instructed in the example code below.

Data Sources

The following is a list of each data source used in our analysis and a description of its provenance and availability:

- Current Population Survey Outgoing Rotation Group data (CPS ORG) for 2018 and 2019: These data were downloaded from the IPUMS website (https://cps.ipums.org/cps/index.shtml).[1] To retrieve data from the website, an individual must follow these steps:
  - Create an account and log in.
  - On the left-hand-side menu, click DATA/BROWSE AND SELECT DATA.
  - Click the "SELECT SAMPLES" button.
  - On the "SELECT SAMPLES" page, toggle the "All Default Samples" checkbox to deselect any preselected samples.
  - Click on the "Basic Monthly" tab and select the months of April through September for 2018 and 2019, then click "SUBMIT SAMPLE SELECTION."
  - Under "SELECT VARIABLES" select the following variables (this can be done using the "A-Z" tab or by simply using the "SEARCH" tab and entering each variable one at a time (some of these will be preselected, as will others that you may remove if desired): AGE, SEX, CPSIDP, EARNWEEK, EARNWT, EDUC, ELIGORG, EMPSTAT, MISH, MONTH, WKSTAT, YEAR.
  - After variable selection is complete, click "VIEW CART" and verify that the described samples and variables are selected.
  - Click "CREATE DATA EXTRACT," enter a description of your extract, and click "SUBMIT EXTRACT."
  - When you are notified that your extract is ready, follow provided link, click "Download .DAT" and save the zipped data extract. Then, under "COMMAND FILES,"

click the "Stata" hyperlink. This will open a new page with computer code for loading these data. Copy this code and paste it into the data cleaning "do file" described below where indicated.

– Unzip the data extract and place this into the "raw" folder, one of multiple folders the user is instructed to create below.

• DMDC administrative data: Access to these data are restricted, so there is no analogous template to follow, but the following variables are needed to generate the weights in the DECI using the provided Stata code below (the exact form and naming conventions for them may vary, but they are described to minimize the potential for confusion or ambiguity):

– period (a date variable in day-month-year format, e.g., "15sep2001")
– svc_cd (a string variable for service, e.g. "A" "N" "F" "M")
– pay_pln_cd (a string variable denoting enlisted, "ME," officer, "MO," or warrant officer "MW")
– age (a numeric age variable in either integer form or in decimal form)
– short_edu_lvl_cd (a string variable with 7 education levels corresponding to "Less than HS or GED," "HS or GED," "Some college," "AA, prof. nursing diploma, or 3-4 years of college but no degree," "BA," "MA+ (incl. 1 or more years of grad school but no degree)," "Unknown").

These data were used in our analysis for comparison but are not required to calculate the DECI:

• Basic pay data: These data were compiled from Table II-1C from the eighth edition (July 2018) of the *Military Compensation Background Papers* (data through 2015; USD[P&R], 2018) and the Congressional Research Service document *Defense Primer: Military Pay Raise* (January 6, 2020).

• Employment Cost Index: These data are derived from a continuous index series from 1976 to 2019 measuring end-of-September 12-month percentage changes (Bureau of Labor Statistics, National Compensation Survey, 2020, Table 9).

Code for Calculating DECI

A few general notes on the code provided below:

• Each section of code comprises the contents of a single Stata "do file," the scripting format used in this software package. These files perform the following functions when run in the following order:
   a. Loading in and preparing the publicly available CPS data.
   b. Imputing missing earnings and checking the quality of the CPS data (e.g., missing cells).
   c. Preparing the military personnel data and generating weights.
   d. Generating a DECI value from these data.

• Note that in the code below and all other subsequent code, the user must substitute a file path and construct a folder structure as directed in the code comments for each do file (Stata comments are either single lines preceded by an *, multiple lines between the /* and */ operators, or in-line following a line of code and preceded by //).

- The execution of this code requires a user to create a specific folder structure, described in the comments to each do file below and the user must substitute a correct file path into the global macro "path" as directed.
- In the three custom programs below used to construct the DECI, an effort has been made to program error messages that will help troubleshoot problems that may be encountered by users of this code, but it is important to read and follow all the instructions throughout the code carefully as well.
- If copying and pasting this code into a set of Stata do files, be careful to assure that long comment lines that are broken to fit on the pages below are corrected in the do file (using backspace to reconstruct broken single lines of commenting).

CPS Data Load-In and Preparation

The following Stata code will generate a cleaned CPS ORG dataset suitable for use in generating a DECI. For this code to work, the zipped (.gz) file from IPUMS (accessed through the procedure described above) must be unzipped and placed into the "raw" data folder created by the user (see folder structure in comments at top of do file below) and the Stata code provided with the zip (as previously mentioned) must be pasted into this do file in the indicated place.

```
* CPS data load-in and preparation
/*
This do file inspects and cleans IPUMS CPS ORG data. Running this code requires
following the instructions for downloading IPUMS data as given in text and
requires creating the following folder structure:
TOP-LEVEL FOLDER NAME: analysis
NEXT-LEVEL FOLDERS (inside "analysis"): raw, working, do, output
SUB-FOLDER INSIDE "output" FOLDER: tables, figures
*/


* global directory path (replace with your path to "analysis" folder)
global path "INSERT YOUR DIRECTORY PATH INSIDE QUOTES"


* other folder macros
global raw "$path/raw"
global work "$path/working"
global do "$path/do"
global output "$path/output"


* enter year of DECI in local macro below
local yr2 2019 // replace this year with DECI year of interest if different
local yr1=(`yr2'-1)


/*
install needed packages (if computer is online, line below will install
"unique" package. otherwise, user will need to download this and follow
```

```
instructions (available online for manually installing Stata packages. If
manually installed, comment out line below by placing an asterisk before it.
*/
ssc install unique, replace all

clear
set more off, perm
set type double

cd "$raw"
*********************************************************************
***** CODE FROM IPUMS DATA EXTRACT FOR LOADING IN CPS DATA ******
*********************************************************************

(ERASE THIS LINE and insert Stata code from IPUMS here)

*********************************************************************

keep if year==`yr1' | year==`yr2'

set seed 123456 // so randomly dropping multiple observations below is
reproducible

* set NIU and missing incwage values to missing
replace earnweek=. if earnweek>=9999.99

/* keep only respondents eligible for the earnings supplement during
months earnings are measured. */
keep if eligorg==1
keep if mish==4 | mish==8
unique cpsidp
duplicates report cpsidp
duplicates tag cpsidp, gen(duplflag)
tab duplflag mish
/* generate number to randomly drop mish=4 or mish=8 for those with 2 obs., then
keep (arbitrarily) lower number of these 2 obs. */
gen randnum=runiform()
egen keeptemp=min(randnum), by(cpsidp)
gen keep=(keeptemp==randnum)
drop if keep==0 & duplflag==1
drop keeptemp randnum keep
duplicates report cpsidp

* explore distribution of missing educ values across years
tab year if educ==1
gen edmiss=(educ==1)
```

```
bys year: tab edmiss
drop if educ==1
drop edmiss

* gen female indicator
gen female=(sex==2)

* gen educ groups
gen hsd=(educ<=71)
gen hsg=(educ==73)
gen sc=(educ>=80 & educ<110)
gen bacc=(educ>=110 & educ<=121)
gen maplus=(educ>=122)

gen educgrp=.
replace educgrp=1 if hsd==1
replace educgrp=2 if hsg==1
replace educgrp=2 if sc==1
replace educgrp=3 if bacc==1
replace educgrp=4 if maplus==1

drop if age<17
* gen age groups
gen age1721=(age>=17 & age<=21)
gen age2226=(age>=22 & age<=26)
gen age2731=(age>=27 & age<=31)
gen age3236=(age>=32 & age<=36)
gen age3741=(age>=37 & age<=41)
gen age4246=(age>=42 & age<=46)
gen age4751=(age>=47 & age<=51)
gen age52plus=(age>=52)

gen agegrp=.
replace agegrp=1 if age1721==1
replace agegrp=2 if age2226==1
replace agegrp=3 if age2731==1
replace agegrp=4 if age3236==1
replace agegrp=5 if age3741==1
replace agegrp=6 if age4246==1
replace agegrp=7 if age4751==1
replace agegrp=8 if age52plus==1

* keep "usually full-time" workers currently working full-time
keep if wkstat<=11

cd "$work"
```

```
save cps_org_april_sept_`yr1'_`yr2'_cleaned, replace
```

CPS Earnings Imputation and Data Quality Check

The following Stata code will further prepare the CPS data for use in DECI generation by imputing any earnings for age-by-education groups without a valid wage measure and assuring there are no missing age-by-education cells.

```
* CPS wage imputation and DECI cell generation
/*
This do file executes wage imputation if required, checks for complete cell
structure, and generates a sequential count of "DECI cells."
Running this code requires creating the following folder structure:
TOP-LEVEL FOLDER NAME: analysis
NEXT-LEVEL FOLDERS (inside "analysis"): raw, working, do, output
SUB-FOLDER INSIDE "output" FOLDER: tables, figures


THIS CODE CREATES A PROGRAM TO PERFORM A FLEXIBLE PROCESS OF IMPUTING EARNINGS
DATA ACCORDING TO USER SPECIFICATIONS! PARAMETERS TO RUN IT MUST BE
ENTERED AFTER THE PROGRAM CODE BELOW. THE LINE "end" DENOTES THE END OF THE
PROGRAM DEFINITION. THERE ARE EXTENSIVE INSTRUCTIONS BELOW GUIDING USER ON HOW
TO SPECIFY THESE PARAMETERS AND EXAMPLE CODE TO RUN THE PROGRAM.
*/


* global directory path (replace with your path to "analysis" folder)
global path "INSERT YOUR DIRECTORY PATH INSIDE QUOTES"

* other folder macros
global raw "$path/raw"
global work "$path/working"
global do "$path/do"
global output "$path/output"

clear
clear matrix
clear mata
set maxvar 12000 // set maxvar to accommodate earnings imputation regression
below

cap program drop cps_deci_gen
program define cps_deci_gen
        syntax [if] [in], [cps_file(string)] [yeart(real 4)] [gender(real 1)]
[output_file_name(string)]


********************************************************************************
*************************** DECI CELL CREATION *********************************
********************************************************************************
```

```
local age_groups 8
local educ_groups 4
local yeartminus1=(`yeart'-1)
local earn_measure earnweek
local cps_weight earnwt
local sex_groups=(`gender'+1)
di `sex_groups'

cd "$work"
use `cps_file', clear
di "year is `yeart'"
di "yeartminus1 is `yeartminus1'"
keep if year==`yeart' | year==`yeartminus1'
keep if `earn_measure'!=.

preserve
* generate dataset to use below in checking for missing age-by-educ-
(by-gender)-by-year cells
cap assert `gender'==1 // expand or not based on gender specification
        if _rc==0 {
                gen obs=1
                collapse (sum)obs, by(agegrp educgrp female year)
                save ageeducgrps_valid, replace
                }
        else {
                gen obs=1
                collapse (sum)obs, by(agegrp educgrp year)
                save ageeducgrps_valid, replace
                }
restore

* GENERATE AND APPEND APPROPRIATE DUMMY DATASET TO FILL ANY EMPTY AGE-ED-GENDER
CELLS W/ZERO PLACEHOLDER VALUES
preserve
clear
local sex_groups=(`gender'+1)
di `sex_groups'
set obs `educ_groups'
gen educgrp=_n
expand `age_groups'
sort educgrp
bys educgrp: gen agegrp=_n
cap assert `gender'==1 // expand or not based on gender specification
        if _rc==0 {
                expand `sex_groups'
                sort educgrp agegrp
```

```
            bys agegrp educgrp: gen female=_n
            replace female=female-1
            }
    local yrcount=(1+`yeartplusn'-`yeart')
    expand `yrcount' // expand years according to whether or not gender specified
        if _rc==0 {
                sort educgrp agegrp female
                bys agegrp educgrp female: gen year=_n
                }
        else {
                sort educgrp agegrp
                bys agegrp educgrp: gen year=_n
                }
    replace year=year-1+`yeart'
    gen ones=0 // assures that appending this dataset won't alter cellsize counts
    cd "$work"


    cap assert `gender'==1 // expand or not based on gender specification
        if _rc==0 {
                merge 1:1 agegrp educgrp female year using ageeducgrps_valid
                keep if _merge==1
                drop _merge
                save dummydatatemp
                tab ones
                global miss_count=r(N) // indicates whether dummy dataset is
                required
                restore
                }
        else {
                merge 1:1 agegrp educgrp year using ageeducgrps_valid
                keep if _merge==1
                drop _merge
                save dummydatatemp
                tab ones
                global miss_count=r(N) // indicates whether dummy dataset is
                required
                restore
                }

    cap assert $miss_count==0
    if _rc!=0 {
        append using dummydatatemp
        di "data has age-by-educ cells with no valid earnings measures"
        * impute wage for cells with no respondents by predicting any missing
        wages
                gen logwage=log(`earn_measure'+1)
```

```
            cap assert `gender'==1 // expand or not based on gender
            specification
                    if _rc==0 {
                            qui reg logwage i.agegrp##i.educgrp##i.female##i.year
                            [pw=`cps_weight']  // use non-parametric regression
                            to predict missing wages
                            }
                    else {
                            qui reg logwage i.agegrp##i.educgrp##i.year [pw=`cps_
weight']  // use non-parametric regression to predict missing wages
                            }
            predict logwagehat, xb
            order logwagehat, after(logwage)
            gen wageimpute=(`earn_measure'==.)
            replace logwage=logwagehat if wageimpute==1
            replace `earn_measure'=int(exp(logwage)) if wageimpute==1
            order `earn_measure', after(wageimpute)
            replace `cps_weight'=1 if wageimpute==1 // adds a unit weight for
            cell-years with only one imputed value (otherwise they will be
            dropped)
            }


* generate decicells
cap assert `gender'==1 // expand or not based on gender specification
        if _rc==0 {
                order agegrp educgrp female age
                sort agegrp educgrp female year
                egen decicell=group(agegrp educgrp female)
                sort decicell year age female
                }
        else {
                order agegrp educgrp age
                sort agegrp educgrp year
                egen decicell=group(agegrp educgrp)
                sort decicell year age
                }


tab agegrp educgrp
tab decicell

save `output_file_name'_`age_groups'`educ_groups'`sex_groups', replace
di "CPS data file has been saved to specified working directory with file name
`output_file_name'_`age_groups'`educ_groups'`sex_groups'"

cap erase dummydatatemp.dta
cap erase ageeducgrps_valid.dta
```

```
end
```

```
* INSTRUCTIONS FOR RUNNING "cps_deci_gen" PROGRAM.
/*
In the line of code below beginning with "cps_deci_gen" (which calls the
program above) enter the following parameters to run the program.
1  In parentheses after "cps_file" enter name of cps output data file created
using the earlier CPS data load-in and preparation code.
2  In parentheses after "yeart" enter year of data (this is year of the desired
DECI).
3  In parentheses after "gender" enter a "1" to implement gender-specific
weighting and a "0" otherwise.
4  Enter the desired output file name prefix. This prefix will be followed by
the age and education group counts used in the data. For example, entering
"cps_org_deci" will yield the final file name "cps_org_deci_84.dta."
*/
```

```
**** cps_deci_gen PROGRAM EXAMPLE ****
cps_deci_gen, cps_file(cps_org_april_sept_2018_2019) yeart(2019)
gender(1) output_file_name(cps_org_deci)
```

Military Weights Data Cleaning and Preparation

The following Stata code will generate a set of military weights for use in the DECI if the user has accessed and prepared a data set as described earlier in this appendix. Extensive details and instructions for using this program are included as comments in the code below.

```
* ADMF DATA CLEANING & DECI WEIGHT GENERATION

/*
This do file cleans administrative military personnel data and generates
weights.
Running this code requires creating the following folder structure:
TOP-LEVEL FOLDER NAME: analysis
NEXT-LEVEL FOLDERS (inside "analysis"): raw, working, do, output
SUB-FOLDER INSIDE "output" FOLDER: tables, figures

THIS CODE CREATES A PROGRAM TO PERFORM A FLEXIBLE PROCESS OF GENERATING MILITARY
WEIGHTS ACCORDING TO USER SPECIFICATIONS! PARAMETERS TO RUN IT MUST BE ENTERED
AFTER THE PROGRAM CODE BELOW. THE LINE "end" DENOTES THE END OF THE PROGRAM.
THERE ARE EXTENSIVE INSTRUCTIONS BELOW GUIDING USER ON HOW TO SPECIFY THESE
PARAMETERS AND EXAMPLES OF CODE TO GENERATE THREE DIFFERENT SETS OF WEIGHTS.
*/

* global directory path (replace with your path to "analysis" folder)
global path "INSERT YOUR DIRECTORY PATH INSIDE QUOTES"
```

```
* other folder macros
global raw "$path/raw"
global work "$path/working"
global do "$path/do"
global output "$path/output"

set more off, perm
set type double
clear

cap program drop admf _ weight _ gen
program define admf _ weight _ gen
* make sure "syntax..." string below is in single line if pasting code
        syntax [if] [in], [admf _ file(string)] [year(real 4)] [gender(real 1)]
[agegrp _ keep _ 1(string)] [agegrp _ keep _ 2(string)] [educgrp _ keep _ 1(string)]
[drop _ 1st _ service _ MO(string)] [drop _ 2nd _ service _ MO(string)] [drop _ 3rd _
service _ MO(string)] [drop _ 4th _ service _ MO(string)] [drop _ 1st _ service _
ME(string)] [drop _ 2nd _ service _ ME(string)] [drop _ 3rd _ service _ ME(string)]
[drop _ 4th _ service _ ME(string)] [output _ file _ suffix(string)]


*************************
**** CLEAN ADMIN DATA ****
*************************
di "clean data"
cd "$raw"
use `admf _ file', clear

keep if pay _ pln _ cd=="ME" | pay _ pln _ cd=="MO" // keep comm officers, enlisted
only

* extract discrete date vars
gen year=year(period)
gen month=month(period)
gen yob=year(pn _ brth _ dt)
gen mob=month(pn _ brth _ dt)
gen yrservstart=year(pay _ entry _ base _ dt)
gen moservstart=month(pay _ entry _ base _ dt)

* keep appropriate time period
cap assert year==`year'
        if _ rc!=0 {
        di "Specified year does not match year in data."
        assert year==`year'
        }
keep if month==9 // only use end-of-year personnel (September)
```

```
* keep personnel from 4 main services
keep if svc_cd=="A" | svc_cd=="F" | svc_cd=="N" | svc_cd=="M"

gen ageint=int(age)
tab ageint
drop if ageint<17 // drop any obs w/erroneous age

* gen age group string var
cap drop age_group // var may or may not be present drop and regenerate
gen str6 age_group=""
replace age_group="17-21" if ageint>=17 & ageint<=21
replace age_group="22-26" if ageint>=22 & ageint<=26
replace age_group="27-31" if ageint>=27 & ageint<=31
replace age_group="32-36" if ageint>=32 & ageint<=36
replace age_group="37-41" if ageint>=37 & ageint<=41
replace age_group="42-46" if ageint>=42 & ageint<=46
replace age_group="47-51" if ageint>=47 & ageint<=51
replace age_group="52-80" if ageint>=52 & ageint<=80

* recode age 80+ into age 52-80 group
replace age_group="52-80" if age_group=="80+"
* drop if age_group==""
egen agegrp=group(age_group)
drop if agegrp==.

* gen numeric var for educ codes
gen int educ=.
replace educ=1 if short_edu_lvl_cd=="1. Less than HS or GED"
replace educ=2 if short_edu_lvl_cd=="2. HS or GED"
replace educ=3 if short_edu_lvl_cd=="3. Some college"
replace educ=4 if short_edu_lvl_cd=="4. AA, prof. nursing diploma, or 3-4
years of college but no degree"
replace educ=5 if short_edu_lvl_cd=="5. BA"
replace educ=6 if short_edu_lvl_cd=="6. MA+ (incl. 1 or more years of grad
school but no degree)"
replace educ=7 if short_edu_lvl_cd=="7. Unknown"

* flag "unknown" ed levels
gen educknown=educ
replace educknown=. if educknown==7
egen educmax=max(educknown), by(ssnscr)
egen educmin=min(educknown), by(ssnscr)
gen edconst_temp=(educmax==educmin)
egen edconst=max(edconst_temp), by(ssnscr)
drop edconst_temp
```

```
gen noeduc=(educmax==. & educmin==.)
egen any_ed_unk=max(educ), by(ssnscr)
replace any_ed_unk=0 if any_ed_unk<7
replace any_ed_unk=1 if any_ed_unk==7

gen female=(pn_sex_cd=="F")

cd "$work"
di "generate weights"
local educ_groups=4

di "implement service and officer/enlisted restrictions as specified"
cap assert `"`drop_1st_service_MO'"'!=""
        if _rc==0 {
                drop if pay_pln_cd=="MO" & svc_cd==`drop_1st_service_MO'"
                di "Officers from `drop_1st_service_MO' dropped"
                }
cap assert `"`drop_2nd_service_MO'"'!=""
        if _rc==0 {
                drop if pay_pln_cd=="MO" & svc_cd==`drop_2nd_service_MO'"
                di "Officers from `drop_2nd_service_MO' dropped"
                }
cap assert `"`drop_3rd_service_MO'"'!=""
        if _rc==0 {
                drop if pay_pln_cd=="MO" & svc_cd==`drop_3rd_service_MO'"
                di "Officers from `drop_3rd_service_MO' dropped"
                }
cap assert `"`drop_4th_service_MO'"'!=""
        if _rc==0 {
                drop if pay_pln_cd=="MO" & svc_cd==`drop_4th_service_MO'"
                di "Officers from `drop_4th_service_MO' dropped"
                }
cap assert `"`drop_1st_service_ME'"'!=""
        if _rc==0 {
                drop if pay_pln_cd=="ME" & svc_cd==`drop_1st_service_ME'"
                di "Enlisted from `drop_1st_service_ME' dropped"
                }
cap assert `"`drop_2nd_service_ME'"'!=""
        if _rc==0 {
                drop if pay_pln_cd=="ME" & svc_cd==`drop_2nd_service_ME'"
                di "Enlisted from `drop_2nd_service_ME' dropped"
                }
cap assert `"`drop_3rd_service_ME'"'!=""
        if _rc==0 {
                drop if pay_pln_cd=="ME" & svc_cd==`drop_3rd_service_ME'"
                di "Enlisted from `drop_3rd_service_ME' dropped"
```

```
                }
cap assert `""`drop_4th_service_ME'"'`!=""
        if _rc==0 {
                drop if pay_pln_cd=="ME" & svc_cd=="`drop_4th_service_ME'"
                di "Enlisted from `drop_4th_service_ME' dropped"
                }


tab svc_cd pay_pln_cd

* education group coding
drop if any_ed_unk==1 // drop indivs with any missing ed

* gen 4-way educ codes (lths, hssc, bacc, maplus)
gen int educgrp=.
replace educgrp=1 if educ==1
replace educgrp=2 if educ==2
replace educgrp=2 if educ==3 | educ==4
replace educgrp=3 if educ==5
replace educgrp=4 if educ==6
tab educgrp

di "cut on specified subDECI age and ed groups"
cap assert `educgrp_keep_1'!=.
        di _rc
        if _rc==0 {
                di "keep educ group `educgrp_keep_1'"
                keep if educgrp==`educgrp_keep_1'
                }
cap assert `agegrp_keep_1'!=.
        if _rc==0 {
                cap assert `agegrp_keep_2'!=.
                    if _rc==0 {
                    keep if agegrp==`agegrp_keep_1' |
agegrp==`agegrp_keep_2'
                    }
                    else {
                    keep if agegrp==`agegrp_keep_1'
                    }
                }
tab agegrp educgrp

preserve
* generate dataset to use below in checking for missing age-by-educ-
(by-gender)-by-year cells
cap assert `gender'==1 // expand or not based on gender specification
        if _rc==0 {
```

```
            gen obs=1
            collapse (sum)obs, by(agegrp educgrp female)
            save ageeducgrps _ valid, replace
            }
    else {
            gen obs=1
            collapse (sum)obs, by(agegrp educgrp)
            save ageeducgrps _ valid, replace
            }
restore

* GENERATE AND APPEND APPROPRIATE DUMMY DATASET TO FILL ANY EMPTY AGE-ED-GENDER
CELLS
preserve
clear
di "dummy dataset generation"
local sex _ groups=(`gender'+1)
set obs `educ _ groups'
gen educgrp= _ n
cap assert `educgrp _ keep _ 1'!=. // use presence of educ group cut value to
replace educ group with appropriate group #
        if _ rc==0 {
                keep if educgrp==`educgrp _ keep _ 1'
                }
expand 8 // need to start with this default number of age groups for age
restrictions to be implemented correctly
sort educgrp
bys educgrp: gen agegrp= _ n
cap assert `agegrp _ keep _ 1'!=.
        if _ rc==0 {
                cap assert `agegrp _ keep _ 2'!=.
                        if _ rc==0 {
                        keep if agegrp==`agegrp _ keep _ 1' |
agegrp==`agegrp _ keep _ 2'
                        }
                        else {
                        keep if agegrp==`agegrp _ keep _ 1'
                        }
                }
cap assert `gender'==1 // expand or not based on gender specification
        if _ rc==0 {
                di `sex _ groups'
                expand `sex _ groups'
                sort educgrp agegrp
                bys agegrp educgrp: gen female= _ n
                replace female=female-1
```

```
            }
cap assert `gender'==1 // sort based on gender specification
        if _rc==0 {
                sort educgrp agegrp female
                bys agegrp educgrp female: gen year=_n
                }
        else {
                sort educgrp agegrp
                bys agegrp educgrp: gen year=_n
                }
replace year=`year'
gen ones=0 // value 0 assures that appending this to "real" dataset won't alter
cellsize counts
tab agegrp educgrp
cap tab female
cd "$work"


* generate decicells
cap assert `gender'==1 // expand or not based on gender specification
        if _rc==0 {
                merge 1:1 agegrp educgrp female using ageeducgrps_valid
                keep if _merge==1
                drop _merge
                save dummydatatemp, replace
                tab ones
                global miss_count=r(N) // use this below to indicate whether
                appending dummy dataset is required
                restore
                }
        else {
                merge 1:1 agegrp educgrp using ageeducgrps_valid
                keep if _merge==1
                drop _merge
                save dummydatatemp, replace
                tab ones
                global miss_count=r(N) // use this below to indicate whether
                appending dummy dataset is required
                restore
                }

cap assert r(mean)==0
        if _rc!=0 {
                append using dummydatatemp
                di "data has missing age-by-educ cells"
                }
```

```
cap erase dummydatatemp.dta

* generate decicells
cap assert `gender'==1 // expand or not based on gender specification
     if _rc==0 {
          order agegrp educgrp female age
          sort agegrp educgrp female year
          egen decicell=group(agegrp educgrp female)
          sort decicell year age female educgrp
          }
     else {
          order agegrp educgrp age
          sort agegrp educgrp year
          egen decicell=group(agegrp educgrp)
          sort decicell year age educgrp
          }
cap gen ones=.
replace ones=1 if ones==. // fills in missing "ones" values from var in "dummy_
dataset_***" above. used to count obs in "cellsize" var
egen cellsize=sum(ones), by(decicell year)

cap assert `gender'==1
     if _rc==0 {
          di "collapse down to set of weights"
          collapse (sum)ones (mean)cellsize agegrp educgrp female,
          by(decicell year)
          replace agegrp=int(agegrp) // these lines address a numerical
          rounding issue affecting future merges
          replace educgrp=int(educgrp)
          }
     else {
          di "collapse down to set of weights"
          collapse (sum)ones (mean)cellsize agegrp educgrp, by(decicell year)
          replace agegrp=int(agegrp) // these lines address a numerical
rounding issue affecting future merges
          replace educgrp=int(educgrp)
          }

egen yeartot=sum(ones), by(year)
gen weight=ones/yeartot
egen sumweight=sum(weight), by(year) // check that weights sum to one
rename cellsize cellsize_admf

cap assert `gender'==1
     if _rc==0 {
          keep year cellsize_admf decicell weight agegrp educgrp female
```

```
                }
        else {
                keep year cellsize_admf decicell weight agegrp educgrp
                }

tab agegrp educgrp
tab decicell

cap assert `agegrp_keep_1'!=.
    if _rc==0 {
            cap assert `agegrp_keep_2'!=.
                if _rc==0 {
                di "Specified age groups kept are `agegrp_keep_1' and
`agegrp_keep_2'"
                local age_groups 2
                }
                else {
                di "Specified age group kept is `agegrp_keep_1'"
                local age_groups 1
                }
            }
    else {
            local age_groups 8
            }
di "Specified number of age groups is `age_groups'"

* make sure wrapped lines below starting w/"di..." are grouped into single lines
if pasting code
di "Specified educ category grouping uses `educ_groups' groups"
cap assert `educgrp_keep_1'!=.
    if _rc==0 {
            di "Specified education group kept is `educgrp_keep_1'"
            }

cap assert `gender'==1
    if _rc==0 di "Gender has been specified"
    if _rc!=0 di "Gender has been omitted"

cap assert `educgrp_keep_1'!=.
    if _rc==0 {
            save admf_weights_`age_groups'`educ_groups'`sex_groups'_
            keep_educgrp_`educgrp_keep_1', replace
            di "Requested weights have been saved to specified working
            directory with file name admf_weights_`age_groups'`educ_
            groups'`sex_groups'_keep_educgrp_`educgrp_keep_1'."
```

```
                    local name "admf _ weights _ `age _ groups'`educ _ groups'`sex _
                    groups' _ keep _ educgrp _ `educgrp _ keep _ 1'"
                    local name2 "admf _ weights _ `age _ groups'`educ _ groups'`sex _
                    groups' _ keep _ educgrp _ `educgrp _ keep _ 1'"
                    }
          else {
                    save admf _ weights _ `age _ groups'`educ _ groups'`sex _ groups',
replace
                    di "Requested weights have been saved to specified working
                    directory with file name admf _ weights _ `age _ groups'`educ _
                    groups'`sex _ groups'."
                    local name "admf _ weights _ `age _ groups'`educ _ groups'`sex _
                    groups'"
                    local name2 "admf _ weights _ `age _ groups'`educ _ groups'`sex _
                    groups'"
                    }
cap assert `agegrp _ keep _ 1'!=.
          if _ rc==0 {
                    cap assert `agegrp _ keep _ 2'!=.
                         if _ rc==0 {
                                   save `name' _ keep _ agegrps _ `agegrp _
                                   keep _ 1'`agegrp _ keep _ 2', replace
                                   cap erase `name'.dta
                                   di "File has been saved to specified working
                                   directory with file name
                                   `name' _ keep _ agegrps _ `agegrp _ keep _ 1'`agegrp _
                                   keep _ 2'."
                                   local name2 `name' _ keep _ agegrps _ `agegrp _
                                   keep _ 1'`agegrp _ keep _ 2'
                         }
                         else {
                                   save `name' _ keep _ agegrp _ `agegrp _ keep _ 1', replace
                                   cap erase `name'.dta
                                   di "File has been saved to specified working
                                   directory with file name
                                   `name' _ keep _ agegrp _ `agegrp _ keep _ 1'."
                                   local name2 `name' _ keep _ agegrp _ `agegrp _ keep _ 1'
                         }
                    }

cap assert `"`output _ file _ suffix'"'==""
          if _ rc!=0 {
                    save `name2' _ `output _ file _ suffix', replace
                    cap erase `name2'.dta
                    di "File has been renamed with file name
`name2' _ `output _ file _ suffix'."
```

```
        }

cap erase dummy_dataset.dta

end
```

* INSTRUCTIONS FOR RUNNING "admf_weight_gen" PROGRAM TO GENERATE MILITARY WEIGHTS FOR DECIs & subgroup DECIs.
/*

In the line of code below beginning with "admf_weight_gen" (which calls the program above) enter the following parameters to run the program. Parameters not needed (the sample restrictions) can be left off or can be entered with nothing in the parentheses.

1 In parentheses after "admf_file" enter name of administrative data file.
2 In parentheses after "year" enter year of data (this is year prior to desired DECI, or year "t-1").
3 In parentheses after "gender" enter a "1" to implement gender-specific weighting and a "0" otherwise.
4 The age of service members included in the sample may be manipulated by selecting up to two specific age groups to generate subgroup DECI weights. This can be accomplished by adding the following options, "agegrp_keep_1()" and "agegrp_keep_2()." The argument for these options is a string variable that will take age group integer values between 1 and 8 with the following breakdown: 1=17-21, 2=22-26, 3=27-31, 4=32-36, 5=37-41, 6=42-46, 7=47-51, and 8=52 and older. Entering one option will keep a single age group, entering both options will keep (and analyze together) two age groups.
5 The education of service members included in the sample may be manipulated by selecting one specific education groups to generate subgroup DECI weights. This can be accomplished by adding the following option, "educgrp_keep_1()." The argument for this options is a string variable that will take education group integer values between 1 and 4 with the following breakdown: 1=less than HS degree, 2=HS degree or some college (including associate's degree), 3=bachelor's degree, 4=master's degree or greater.
        NOTE THAT THESE AGE AND EDUCATION OPTIONS CAN BE USED TOGETHER AND WILL GENERATE THE SUBGROUP REPRESENTING THE INTERSECTION OF THE CHOSEN AGE AND EDUCATION GROUPS. THE FILE WILL OUTPUT A NAME WITH THESE OPTIONS REFLECTED IN IT.
6 In parentheses after drop_1st_service_MO enter service code for any desired officer data cut. These codes are Army: "A", Navy: "N", Air Force "F", Marine Corps "M". This is similar for 2nd, 3rd, and 4th entries and also for 1st_service_ME for enlisted as well as 2nd, 3rd, 4th. NOTE THAT THESE SAMPLE RESTRICTIONS WILL NOT AUTOMATICALLY BE REFLECTED IN THE RESULTING WEIGHT FILE NAME. TO MAKE CLEAR WHAT SERVICES HAVE BEEN DROPPED/INCLUDED, USE THE FILE NAME SUFFIX OPTION DESCRIBED IN 6.
7 If desired, a suffix for the weight file created may be added in the parentheses after "output_file_suffix." For example, if a sub-DECI for

officers only is desired, user can drop all 4 enlisted services and then add
"output_file_suffix(officers)" at the end of the option string.

The code below contains three examples of weight creation for an overall
DECI and for two subgroup DECIs, one for only service members ages 27-36
with a bachelors degree, and one for only enlisted personnel. Note that the
administrative data to be used must contain the variables described in the
text.

The files generated by this program will have the following naming conventions:
     a all will begin with admf_weights_
     b all will have a series of numbers representing, respectively the age
       and education groups specified (these will be followed by an additional
       1 to denote pooled gender). thus, a set of weights for all age and
       education groups will be called "admf_weights_841."
     c weights for subgroup DECIs will have age and
       education subgroup choices reflected in the name in the following
       way: if the user chooses, for example age groups 3 and 4 and
       education group 3 (as in the example below), the resulting filename will
       be "admf_weights_211_keep_agegrps_34_keep_educgrp_3."
*/

```
**** admf_weight_gen PROGRAM EXAMPLES ****
* Subgroup DECI: all services, officer and enlisted, ages 27-36 with bachelor's
degree
admf_weight_gen, admf_file(admf_dummy_raw_data) year(2018) gender(0)
agegrp_keep_1(3) agegrp_keep_2(4) educgrp_keep_1(3)

* Subgroup DECI: all services, enlisted only (i.e., drop officers from all 4
services)
admf_weight_gen, admf_file(admf_dummy_raw_data) year(2018) gender(0)
drop_1st_service_MO(A) drop_2nd_service_MO(N) drop_3rd_service_MO(F)
drop_4th_service_MO(M) output_file_suffix(enlisted)

* Full DECI: all services, both officer and enlisted, all age and education
groups
admf_weight_gen, admf_file(admf_dummy_raw_data) year(2018) gender(1)
```

Defense Employment Cost Index (DECI) Generation Code

The following Stata code will generate a DECI or subgroup DECI measure as specified by the user. This code combines the CPS data and the military weight data generated by the prior code. As with the other code, there are extensive instructions and other details in the comments of the code itself.

```
* PROGRAM TO GENERATE DECI
```

```
/*
This do file defines and executes a program to generate a DECI or subgroup DECI
as specified using a CPS data file that is the output of executing the Stata
code "CPS BASIC 1982-2018 cleaning" and "CPS wage imputation and DECI cell
generation" (defined previously in this document) and a military weight
file prepared by  running the Stata code "ADMF data cleaning & DECI weight
generation."

Running this code requires defining a global working directory path, as
instructed below and then creating the following folder structure:
TOP-LEVEL FOLDER NAME: analysis
NEXT-LEVEL FOLDERS (inside "analysis"): raw, working, do, output
SUB-FOLDER INSIDE "output" FOLDER: tables, figures

THIS CODE CREATES A PROGRAM TO GENERATE A DECI! PARAMETERS TO RUN IT MUST BE
ENTERED AFTER THE PROGRAM CODE BELOW. THE LINE "end" DENOTES THE END OF THE
PROGRAM DEFINITION. THERE ARE EXTENSIVE INSTRUCTIONS BELOW GUIDING USER ON HOW
TO SPECIFY THESE PARAMETERS AND EXAMPLE CODE TO RUN THE PROGRAM.
*/


* global directory path (replace with your path to "analysis" folder)
global path "INSERT YOUR DIRECTORY PATH INSIDE QUOTES"

* other folder macros
global raw "$path/raw"
global work "$path/working"
global do "$path/do"
global output "$path/output"

clear
set more off, perm
set type double

cap program drop deci_gen
program define deci_gen
        syntax [if] [in], [weight_file(string)] [cps_deci_
        file(string)] [year(real 4)] [gender(real 1)] [prior_deci_value(real 6)]
        [output_file_name(string)]


*******************************************************************
************* CREATE DECI - CHAINED LASPEYRES INDEX *************
*******************************************************************
local yrt=`year' // these locals for use in file naming/calling below
local yrtmin1=(`yrt'-1)
local earn_measure earnweek
local cps_weight earnwt
```

```
di "year is `year'"
di "yrtminus1 if `yrtmin1'"

cd "$work"
use `weight_file', clear
cap assert year==`yrtmin1'
        di _rc
        if _rc!=0 {
                di "Year of weight file is not equal to DECI year t minus 1, as
                required. Check provided data and user-specified parameters."
                assert year==`yrtmin1'
                }
        else {
                save `weight_file'_temp, replace

* generate file to match appropriate CPS microdata based on age-educ composition
of weight data.
cap assert `gender'==1 // select group decicell ingredients based on gender
specification
        if _rc==0 {
                duplicates drop agegrp educgrp female, force
                keep agegrp educgrp female decicell
                save decicellmap, replace
                }
        else {
                duplicates drop agegrp educgrp, force
                keep agegrp educgrp decicell
                save decicellmap, replace
                }

                ****** GENERATE DENOMINATOR OF INDEX ******
                use `cps_deci_file', clear
                cap assert year==`yrtmin1' | year==`yrt'
                if _rc!=0 di "CPS data does not contain required years"
                assert year==`yrtmin1' | year==`yrt'
                * use age-educ structure of weight data to select appropriate CPS
data
                drop decicell
                cap assert `gender'==1 // select group decicell ingredients based
                on gender specification
                        if _rc==0 {
                                merge m:1 agegrp educgrp female using decicellmap //
                                this reconfigures decicells to match weight file
                                characteristics
                                keep if _merge==3
```

```
                }
        else {
                merge m:1 agegrp educgrp using decicellmap // this
                reconfigures decicells to match weight file
                characteristics
                keep if _merge==3
                }


* tabulate and display decicell count, age group count and educ
group count
tab decicell
tab agegrp
local age_groups `r(r)'
tab educgrp
local educ_groups `r(r)'
di "number of age groups is `age_groups'"
di "educ groups is `educ_groups'"


* generate weighted earnings avg within cell-years
collapse `earn_measure' agegrp educgrp [pw=`cps_weight'],
by(decicell year)


* merge with military weights
merge 1:1 decicell year using `weight_file'_temp
keep if _merge==3
* check that weights sum to one.
egen wt_sum=sum(weight), by(year)
summ wt_sum


*collapse down using military weights
collapse `earn_measure' [pw=weight], by(year)
rename `earn_measure' `earn_measure'_denom
save deci_denom_temp, replace


***** GEN. NUMERATOR OF INDEX & MERGE ******
use `cps_deci_file', clear
drop decicell
cap assert `gender'==1 // select group decicell ingredients based
on gender specification
        if _rc==0 {
                merge m:1 agegrp educgrp female using decicellmap //
                this reconfigures decicells to match weight file
                characteristics
                keep if _merge==3
                }
        else {
```

```
                    merge m:1 agegrp educgrp using decicellmap // this
                    reconfigures decicells to match weight file
                    characteristics
                    keep if _merge==3
                    }
cap erase decicellmap.dta

* generate weighted earnings avg within cell-years
collapse `earn_measure' agegrp educgrp [pw=`cps_weight'],
by(decicell year)
replace year=year-1 // adjustment so current year's wages will
merge to prior year's military composition (weights)

* merge with military weights
merge 1:1 decicell year using `weight_file'_temp
keep if _merge==3
egen wt_sum=sum(weight), by(year) // check that all weights sum to
~1.

*collapse down using military weights
collapse `earn_measure' [pw=weight], by(year)
rename `earn_measure' `earn_measure'_numer
save deci_numer_temp, replace

* merge denominator values and create index
merge 1:1 year using deci_denom_temp // merge denominator
observations
drop _merge
replace year=year+1 // return year value to "year t"
gen deci_value_`yrtmin1' = `prior_deci_value'
gen deci_ratio_`yrt' = (`earn_measure'_numer/`earn_measure'_
denom)
gen deci_pct_chg_`yrt' = (deci_ratio_`yrt'-1)*100
gen deci_value_`yrt' = (`earn_measure'_numer/`earn_measure'_
denom)*`prior_deci_value' // gen decivalue
drop earnweek*
preserve
tostring(deci_value_`yrtmin1'), replace force
tostring(deci_ratio_`yrt'), replace force
tostring(deci_pct_chg_`yrt'), replace force
tostring(deci_value_`yrt'), replace force
cd "$output"
export excel using "`output_file_name'_`yrt'", sheetreplace
firstrow(variables)
restore
```

```
            cap erase deci_denom_temp.dta
            cap erase deci_numer_temp.dta
      }

end


/*
The program "deci_gen" uses two data inputs, a file generated by the program
"cps_deci_gen" and a file generated by the program "admf_weight_gen" to
generate a deci for age and education groups, as well as a prior index value
specified by the user. The inputs for this program are:
1) weight_file: string var specifying admf weight file (data file output from
"admf_weight_gen")
2) cps_deci_file: string var specifying cps file (data file output from
"cps_deci_gen")
3) year: real var specifying year for which deci is to be generated (4 digits)
4) gender: enter 1 for gender-specific weighting or 0 for non-gender
specific-weighting (e.g., when estimating subgroup DECIs with smaller numbers of
observations, as discussed in text).
4) prior_deci_value: real var specifying prior index value of deci to use in
calculation (up to 6 digits, can accommodate decimal value, e.g., 162.46)
5) output_file_name: string var specifying the prefix file name to use for
output file (e.g., "deci_all_services"). Added suffix will be user-specified
year.

Note that for generating subgroup decis, it is the weight file where these
choices are made. This program simply uses the weights generated for
whatever group or subgroup analysis is desired and generates the corresponding
deci.
The output from this program is an excel file with the year specified, the
prior deci value specified, the new deci ratio (see text for formula), and the
new deci value.
*/


* deci_gen PROGRAM EXAMPLE
deci_gen, weight_file(cps_org_deci_842) cps_deci_file(cps_
org_deci_842) year(2019) gender(1) prior_deci_value(100.00)
output_file_name(deci_output_2019)
```

Data Points for Figures in Chapter Four and Appendix D

This appendix gives the data points for the figures presented in Chapter Four. This follows the example of Hosek 1992, in which the figures showing the DECI over time were supplemented by tables in Appendix D of Hosek 1992. We found these tables invaluable in our effort to replicate the earlier DECI computations, and we hope that future analysts will find these tables similarly valuable.

Table F.1
Data Points for Figure 4.1

Year	Active Duty Military				Civilian Labor Force			
	Ages 17–21	Ages 22–26	Ages 27–36	Ages 37+	Ages 17–21	Ages 22–26	Ages 27–36	Ages 37+
1982	32	63	90	100	9	26	56	100
1983	31	61	89	100	9	28	59	100
1984	31	61	89	100	9	27	58	100
1985	30	60	88	100	8	24	55	100
1986	29	60	88	100	8	24	55	100
1987	29	59	88	100	8	25	57	100
1988	29	58	87	100	8	25	57	100
1989	29	58	87	100	8	24	56	100
1990	29	57	86	100	8	24	56	100
1991	27	55	86	100	7	23	55	100
1992	26	55	85	100	7	22	54	100
1993	25	55	85	100	7	21	52	100
1994	25	55	85	100	6	20	51	100
1995	28	59	91	100	6	19	49	100
1996	27	59	90	100	6	19	48	100
1997	28	58	89	100	7	21	50	100
1998	28	58	88	100	7	20	49	100
1999	29	58	87	100	7	20	48	100
2000	31	59	87	100	7	20	48	100
2001	31	60	87	100	7	20	46	100
2002	29	60	87	100	6	19	45	100
2003	27	59	87	100	6	19	45	100
2004	26	59	87	100	5	18	45	100
2005	26	59	87	100	6	19	44	100
2006	25	58	87	100	6	19	44	100
2007	24	58	87	100	6	19	44	100
2008	24	57	87	100	5	18	43	100
2009	22	56	86	100	4	17	42	100
2010	21	54	86	100	4	16	42	100
2011	20	53	86	100	4	16	41	100
2012	20	52	85	100	4	16	41	100
2013	21	52	85	100	4	16	42	100
2014	22	52	85	100	4	16	41	100
2015	23	53	86	100	4	16	41	100
2016	23	54	86	100	4	17	43	100
2017	24	54	86	100	4	17	42	100
2018	24	55	86	100	4	16	42	100
2019					4	15	41	100

Table F.2
Data Points for Figure 4.2

	Active Duty Military				Civilian Labor Force				
Year	Less than high school	High school diploma	Some college/ associate's degree	Bachelor's degree	Less than high school	High school diploma	Some college/ associate's degree	Bachelor's degree	Master's degree plus
1982	10	77	85	95	17	59	78	94	100
1983	7	76	85	95	16	58	78	94	100
1984	5	75	84	94	16	57	77	94	100
1985	4	75	84	94	15	56	77	93	100
1986	4	75	84	94	14	56	77	94	100
1987	4	75	84	94	15	56	77	94	100
1988	10	82	84	94	15	55	76	93	100
1989	9	82	84	94	14	54	76	93	100
1990	8	82	84	94	14	54	76	93	100
1991	8	81	84	94	13	53	75	93	100
1992	6	80	83	93	12	48	75	92	100
1993	6	79	83	93	12	47	75	92	100
1994	5	78	82	93	11	46	74	92	100
1995	4	75	79	93	11	45	74	92	100
1996	4	75	80	93	11	45	74	92	100
1997	4	76	80	93	12	46	74	92	100
1998	6	77	81	93	12	45	73	92	100
1999	10	78	82	93	12	44	73	91	100
2000	5	79	83	93	12	45	73	92	100
2001	7	79	83	93	11	43	72	91	100
2002	7	79	83	93	11	43	71	91	100
2003	7	79	83	93	11	42	71	91	100
2004	7	79	83	93	11	43	71	91	100
2005	7	78	83	93	12	43	71	91	100
2006	7	78	83	94	11	42	71	90	100
2007	8	79	83	94	11	41	69	90	100
2008	9	79	84	94	9	39	68	90	100
2009	9	78	83	94	9	39	67	89	100
2010	8	77	83	94	9	38	66	89	100
2011	6	76	83	93	8	37	66	88	100
2012	5	73	82	93	8	36	65	88	100
2013	5	72	81	93	8	36	64	88	100
2014	4	71	80	93	8	36	64	88	100
2015	4	71	80	93	8	35	64	88	100
2016	4	70	80	92	8	34	63	87	100
2017	4	70	79	92	7	35	62	87	100
2018	4	70	79	92	7	34	61	86	100
2019					7	33	60	86	100

Table F.3
Data Points for Figure 4.3

Date	Hires	Quits	Date	Hires	Quits
2002-01-01	4,787	2,631	2010-01-01	4,023	1,810
2002-04-01	4,885	2,566	2010-04-01	4,230	1,874
2002-07-01	4,899	2,556	2010-07-01	4,081	1,847
2002-10-01	4,828	2,435	2010-10-01	4,218	1,899
2003-01-01	4,747	2,430	2011-01-01	4,220	1,938
2003-04-01	4,653	2,300	2011-04-01	4,308	1,924
2003-07-01	4,701	2,280	2011-07-01	4,329	2,026
2003-10-01	4,884	2,417	2011-10-01	4,364	2,002
2004-01-01	4,972	2,457	2012-01-01	4,531	2,106
2004-04-01	5,046	2,545	2012-04-01	4,454	2,142
2004-07-01	4,989	2,586	2012-07-01	4,360	2,037
2004-10-01	5,161	2,674	2012-10-01	4,443	2,051
2005-01-01	5,250	2,707	2013-01-01	4,476	2,227
2005-04-01	5,336	2,756	2013-04-01	4,558	2,246
2005-07-01	5,410	2,919	2013-07-01	4,672	2,335
2005-10-01	5,254	2,864	2013-10-01	4,585	2,346
2006-01-01	5,423	2,948	2014-01-01	4,695	2,385
2006-04-01	5,389	2,944	2014-04-01	4,871	2,489
2006-07-01	5,411	3,005	2014-07-01	4,989	2,646
2006-10-01	5,361	3,001	2014-10-01	5,094	2,617
2007-01-01	5,355	2,955	2015-01-01	5,102	2,746
2007-04-01	5,363	2,946	2015-04-01	5,166	2,739
2007-07-01	5,306	2,893	2015-07-01	5,204	2,821
2007-10-01	5,231	2,823	2015-10-01	5,403	2,906
2008-01-01	5,024	2,790	2016-01-01	5,359	2,924
2008-04-01	4,844	2,682	2016-04-01	5,273	2,986
2008-07-01	4,606	2,475	2016-07-01	5,340	3,012
2008-10-01	4,333	2,194	2016-10-01	5,302	3,025
2009-01-01	4,005	1,917	2017-01-01	5,410	3,126
2009-04-01	3,762	1,694	2017-04-01	5,471	3,106
2009-07-01	3,870	1,627	2017-07-01	5,484	3,164
2009-10-01	3,955	1,741	2017-10-01	5,520	3,185

Table F.4
Data Points for Figures 4.4, 4.5, 4.10, and 4.13

Year	CPS All	Less Than High School	High School Diploma	Some College/ Associate's Degree	Bachelor's Degree	Master's Degree Plus
1982	100.00	100.00	100.00	100.00	100.00	100.00
1983	101.84	101.43	101.40	101.01	101.03	105.01
1984	107.64	105.99	106.04	107.27	107.86	109.79
1985	114.02	109.34	111.22	113.70	114.18	117.42
1986	117.17	111.03	113.88	115.93	118.37	122.63
1987	119.91	112.86	116.56	119.07	121.26	126.62
1988	123.46	115.56	119.09	122.67	124.86	128.81
1989	133.63	119.27	125.11	130.64	137.57	151.90
1990	137.61	121.31	127.99	136.77	143.92	154.51
1991	144.39	124.49	133.16	142.33	148.59	166.80
1992	147.74	125.90	136.31	139.31	153.98	166.63
1993	152.40	128.44	141.05	143.97	157.10	171.08
1994	159.27	127.00	146.83	148.85	164.00	184.94
1995	165.71	130.53	151.10	153.56	171.03	193.98
1996	168.64	134.38	154.80	157.67	173.04	193.31
1997	171.56	134.33	158.14	159.71	177.10	196.70
1998	182.72	141.42	164.66	169.64	193.53	211.07
1999	191.64	144.78	171.81	174.35	202.80	222.00
2000	199.19	148.87	177.13	182.58	213.68	231.72
2001	208.47	158.27	185.14	190.02	221.35	237.00
2002	214.45	159.38	190.40	195.62	224.36	240.27
2003	219.27	164.40	194.76	197.75	229.37	245.50
2004	224.50	168.66	201.84	204.16	228.44	253.25
2005	227.48	172.30	204.15	203.71	237.77	254.28
2006	235.66	177.29	210.04	210.63	246.74	262.77
2007	245.17	185.86	214.11	219.78	255.65	267.44
2008	256.35	192.13	223.06	226.78	260.84	281.39
2009	260.87	193.88	225.24	224.64	266.08	285.79
2010	263.48	190.93	226.02	228.56	265.23	286.13
2011	267.80	196.96	231.67	231.86	264.58	286.16
2012	274.11	200.25	236.35	233.09	271.51	293.20
2013	278.21	207.14	236.50	235.98	277.71	293.84
2014	281.57	203.69	244.01	242.96	277.90	289.63
2015	287.53	214.20	247.80	246.27	286.74	293.53
2016	296.58	217.51	250.03	248.96	293.37	303.57
2017	305.62	232.44	261.30	257.57	296.61	309.10
2018	314.26	235.04	266.92	264.25	305.58	312.95
2019	328.91	255.39	276.45	274.71	315.46	327.30

Table F.5
Data Points for Figures 4.6 and 4.10

Year	DECI	ECI	BPI
1982	100.00	100.00	100.00
1983	100.40	106.90	104.00
1984	104.47	112.14	104.00
1985	108.90	116.85	108.16
1986	110.86	122.34	115.73
1987	114.96	126.38	115.73
1988	117.01	130.42	119.20
1989	124.77	135.12	121.59
1990	128.46	141.06	126.57
1991	131.64	146.99	131.13
1992	132.32	152.28	136.51
1993	135.62	156.39	142.24
1994	139.56	161.39	147.50
1995	144.44	166.07	150.75
1996	147.25	170.56	154.67
1997	152.08	176.19	158.38
1998	162.01	182.53	163.13
1999	168.60	190.38	167.70
2000	179.19	196.28	173.73
2001	181.30	204.52	184.51
2002	182.87	211.89	192.07
2003	185.88	218.45	205.32
2004	189.61	225.01	214.97
2005	191.90	230.86	224.00
2006	199.32	235.94	231.84
2007	205.22	243.02	239.03
2008	212.78	251.28	245.48
2009	208.19	258.56	254.07
2010	209.64	262.18	263.98
2011	212.33	266.38	272.96
2012	213.89	270.91	276.78
2013	216.25	275.78	281.21
2014	222.56	280.75	285.99
2015	224.46	287.21	288.85
2016	229.88	293.24	291.74
2017	237.88	300.27	295.53
2018	245.22	308.08	301.74
2019	256.21	317.63	308.98

Table F.6
Data Points for Figure 4.7

Year	DECI Earnings Gap	ECI Earnings Gap
1982		
1983	0.035	−0.028
1984	−0.004	−0.078
1985	−0.007	−0.080
1986	0.042	−0.057
1987	0.007	−0.092
1988	0.018	−0.094
1989	−0.026	−0.111
1990	−0.015	−0.114
1991	−0.004	−0.121
1992	0.031	−0.116
1993	0.047	−0.099
1994	0.054	−0.094
1995	0.042	−0.102
1996	0.048	−0.103
1997	0.040	−0.112
1998	0.007	−0.119
1999	−0.005	−0.135
2000	−0.031	−0.130
2001	0.017	−0.108
2002	0.048	−0.103
2003	0.095	−0.064
2004	0.118	−0.047
2005	0.143	−0.031
2006	0.140	−0.018
2007	0.141	−0.017
2008	0.133	−0.024
2009	0.181	−0.018
2010	0.206	0.007
2011	0.222	0.024
2012	0.227	0.021
2013	0.231	0.019
2014	0.222	0.018
2015	0.223	0.006
2016	0.212	−0.005
2017	0.195	−0.016
2018	0.187	−0.021
2019	0.171	−0.028

Table F.7
Data Points for Figure 4.8

Year	DECI	ECI	BPI
1982			
1983			
1984			
1985			
1986			
1987			
1988			
1989			
1990			
1991			
1992			
1993			
1994			
1995			
1996			
1997			
1998			
1999			
2000			
2001			
2002			
2003			
2004			
2005			
2006			
2007			
2008			
2009			
2010	100.00	100.00	100.00
2011	101.28	103.40	101.60
2012	102.03	104.85	103.33
2013	103.15	106.53	105.19
2014	106.16	108.34	107.08
2015	107.07	109.42	109.54
2016	109.65	110.51	111.84
2017	113.47	111.95	114.53
2018	116.97	114.30	117.51
2019	122.21	117.04	121.15

Table F.8
Data Points for Figure 4.9

Year	DECI Earnings Gap	ECI Earnings Gap
1982		
1983		
1984		
1985		
1986		
1987		
1988		
1989		
1990		
1991		
1992		
1993		
1994		
1995		
1996		
1997		
1998		
1999		
2000		
2001		
2002		
2003		
2004		
2005		
2006		
2007		
2008		
2009		
2010	0.000	0.000
2011	0.017	0.020
2012	0.015	0.027
2013	0.013	0.032
2014	0.012	0.020
2015	−0.001	0.021
2016	−0.012	0.008
2017	−0.023	−0.014
2018	−0.028	−0.023
2019	−0.035	−0.044

Table F.9
Data Points for Figure 4.11

Year	CPS Earnings Gap
1982	
1983	0.021
1984	−0.035
1985	−0.054
1986	−0.012
1987	−0.036
1988	−0.036
1989	−0.099
1990	−0.087
1991	−0.101
1992	
1993	
1994	
1995	
1996	
1997	
1998	
1999	
2000	
2001	
2002	
2003	
2004	
2005	
2006	
2007	
2008	
2009	
2010	
2011	
2012	
2013	
2014	
2015	
2016	
2017	
2018	
2019	

Table F.10
Data Points for Figure 4.12

Year	Ages 17–21	Ages 22–26	Ages 27–36	Ages 37+
1982	100	100	100	100
1983	99	101	102	104
1984	103	104	108	110
1985	107	109	113	115
1986	108	110	116	119
1987	114	115	119	122
1988	116	119	122	126
1989	124	124	130	137
1990	129	130	133	142
1991	131	132	138	149
1992	130	133	141	152
1993	131	135	146	156
1994	136	138	148	165
1995	141	143	155	169
1996	146	144	158	172
1997	153	153	162	175
1998	165	160	173	186
1999	175	166	178	196
2000	183	177	190	200
2001	187	184	197	210
2002	189	190	200	216
2003	194	186	204	221
2004	197	196	208	226
2005	198	197	211	229
2006	212	202	219	238
2007	218	212	224	248
2008	221	225	235	257
2009	214	223	235	261
2010	220	224	238	263
2011	230	226	240	268
2012	219	227	248	274
2013	231	227	250	280
2014	244	241	252	281
2015	251	240	257	289
2016	252	247	270	298
2017	264	263	278	305
2018	282	265	286	314
2019	288	287	301	324

Table F.11
Data Points for Figure 4.14

Year	Males	Females
1982	100	100
1983	101	104
1984	106	111
1985	112	118
1986	116	122
1987	118	127
1988	121	132
1989	131	142
1990	134	149
1991	140	157
1992	143	162
1993	146	169
1994	153	173
1995	160	179
1996	162	184
1997	164	191
1998	176	199
1999	185	209
2000	192	217
2001	202	228
2002	205	239
2003	211	243
2004	215	250
2005	215	257
2006	224	265
2007	233	276
2008	244	288
2009	248	296
2010	248	303
2011	251	310
2012	259	314
2013	261	321
2014	264	325
2015	270	331
2016	278	344
2017	285	356
2018	294	365
2019	308	382

Table F.12
Data Points for Figures 4.16–4.18

Year	BPI/Enlisted DECI	BPI/Officer DECI	BPI/ECI	High-Quality Accessions (DoD)	High-Quality Accessions (Army omitted)	Enlisted Continuation (4 YOS)	Officer Continuation (8 YOS)
1982	1.000	1.000	1.000	45.00	48.53	66.23	92.94
1983	1.041	1.009	0.973	52.00	56.85	65.53	93.77
1984	1.004	0.950	0.927	53.00	58.12	65.65	92.68
1985	1.003	0.944	0.926	56.00	57.69	66.35	91.90
1986	1.057	0.970	0.946	57.00	58.87	68.93	91.84
1987	1.019	0.940	0.916	60.00	61.97	63.90	91.53
1988	1.034	0.941	0.914	61.00	61.15	65.90	91.97
1989	1.002	0.850	0.900	58.00	59.56	62.81	89.57
1990	1.013	0.855	0.897	64.00	64.27	62.10	87.80
1991	1.030	0.851	0.892	72.00	68.55	65.30	88.80
1992	1.068	0.874	0.896	74.00	72.35	57.10	85.50
1993	1.085	0.888	0.910	67.00	68.19	61.10	83.80
1994	1.096	0.881	0.914	68.00	68.78	62.30	90.40
1995	1.084	0.872	0.908	67.00	67.63	64.30	88.10
1996	1.088	0.878	0.907	65.00	67.35	64.10	90.30
1997	1.078	0.879	0.899	63.00	65.78	61.00	90.70
1998	1.046	0.837	0.894	63.00	65.31	59.10	90.00
1999	1.038	0.815	0.881	59.00	62.19	58.00	90.50
2000	1.024	0.791	0.885	57.00	60.75	59.90	90.10
2001	1.042	0.822	0.902	59.00	61.51	62.70	91.40
2002	1.076	0.848	0.906	62.30	64.85	67.50	93.10
2003	1.134	0.891	0.940	65.00	68.47	71.00	92.50
2004	1.162	0.916	0.955	67.00	70.81	65.30	91.50
2005	1.202	0.930	0.970	64.00	68.05	64.40	90.60
2006	1.190	0.947	0.983	62.00	69.21	66.70	90.60
2007	1.195	0.939	0.984	59.00	67.62	68.00	86.40
2008	1.185	0.923	0.977	59.00	68.48	71.60	91.40
2009	1.265	0.938	0.983	64.00	73.00	72.00	92.80
2010	1.299	0.988	1.007	70.00	78.16	72.30	94.00
2011	1.324	1.018	1.025	77.00	84.03	70.80	92.20
2012	1.335	1.015	1.022	76.40	86.59	67.50	91.80
2013	1.341	1.023	1.020	73.70	82.60	67.70	92.70
2014	1.315	1.042	1.019	73.48	83.93	68.50	90.30
2015	1.320	1.032	1.006	73.71	83.91	70.00	88.80
2016	1.305	1.010	0.995	71.04	80.24	70.40	91.10
2017	1.270	1.011	0.984	69.87	78.54	71.50	90.60
2018	1.255	1.011	0.979	68.95		72.60	91.00
2019	1.233	0.988	0.973	66.40			

Table F.13
Data Points for Figures D.2 and D.9

Year	Figure D.2		Figure D.9	
	DECI (ORG)	DECI (ASEC)	DECI (SOFS)	DECI (ADMF)
1982	100.00	100.00		
1983	100.40	102.77		
1984	104.47	105.49		
1985	108.90	110.73		
1986	110.86	115.80		
1987	114.96	119.58		
1988	117.01	123.40		
1989	124.77	129.33		
1990	128.46	134.37		
1991	131.64	136.53		
1992	132.32	138.95		
1993	135.62	141.04		
1994	139.56	144.80		
1995	144.44	149.98		
1996	147.25	160.96		
1997	152.08	168.52		
1998	162.01	174.02		
1999	168.60	182.82		
2000	179.19	185.46		
2001	181.30	200.44		
2002	182.87	207.77	100.00	100.00
2003	185.88	209.86		101.51
2004	189.61	209.73	103.12	103.61
2005	191.90	214.09		104.80
2006	199.32	218.94	107.87	108.91
2007	205.22	228.81		112.11
2008	212.78	232.23	114.47	116.34
2009	208.19	235.37		113.83
2010	209.64	233.32	114.03	114.63
2011	212.33	229.84		116.06
2012	213.89	235.50	115.56	116.94
2013	216.25	238.93		118.23
2014	222.56	248.61	120.19	121.68
2015	224.46	252.17		122.72
2016	229.88	260.57	124.23	125.68
2017	237.88	271.81		130.05
2018	245.22	274.88	131.72	134.07
2019	256.21	283.91		

Table F.14
Data Points for Figures D.3–D.5

	Figure D.3		Figures D.4 and D.5		
Year	ECI Gap	ECI Gap (Hosek 92)	DECI Gap (ASEC)	DECI Gap (ORG)	DECI Gap (Hosek 92)
1983	−0.028	−0.028	0.012	−0.008	0.035
1984	−0.078	−0.078	−0.014	−0.031	−0.004
1985	−0.080	−0.078	−0.024	−0.032	−0.007
1986	−0.057	−0.053	−0.001	−0.008	0.042
1987	−0.092	−0.087	−0.033	−0.032	0.007
1988	−0.094	−0.091	−0.035	−0.046	0.018
1989	−0.111	−0.111	−0.064	−0.065	−0.026
1990	−0.114	−0.112	−0.062	−0.068	−0.015
1991	−0.121	−0.119	−0.041	−0.047	−0.004

Table F.15
Data Points for Figure D.6

	Panel A (Baseline Year 1982)		Panel B (Baseline Year 2010)	
Year	DECI Gap High School Graduate/ Some College/ Associate's Degree	DECI Gap Bachelor's Degree	DECI Gap High School Graduate/ Some College/ Associate's Degree	DECI Gap Bachelor's Degree
1982	0.00	0.00		
1983	0.03	0.03		
1984	−0.01	−0.04		
1985	−0.01	−0.05		
1986	0.04	−0.02		
1987	0.00	−0.06		
1988	0.01	−0.07		
1989	−0.02	−0.13		
1990	−0.01	−0.13		
1991	0.00	−0.11		
1992	0.03	−0.11		
1993	0.05	−0.09		
1994	0.06	−0.08		
1995	0.05	−0.09		
1996	0.05	−0.08		
1997	0.04	−0.10		
1998	0.02	−0.15		
1999	0.01	−0.16		
2000	0.00	−0.21		
2001	0.02	−0.17		
2002	0.04	−0.14		
2003	0.09	−0.06		
2004	0.12	−0.04		
2005	0.14	−0.02		
2006	0.14	−0.02		
2007	0.14	−0.02		
2008	0.13	−0.02		
2009	0.19	−0.01		
2010	0.20	0.05	0.00	0.00
2011	0.22	0.07	0.02	0.00
2012	0.23	0.06	0.03	0.00
2013	0.23	0.07	0.04	0.00
2014	0.21	0.07	0.01	0.00
2015	0.22	0.06	0.02	0.00
2016	0.21	0.03	0.01	0.00
2017	0.19	0.05	−0.02	0.00
2018	0.18	0.04	−0.04	0.00
2019	0.16	0.02	−0.05	0.00

Table F.16
Data Points for Figure D.7

Year	Panel A (Baseline Year 1982)		Panel B (Baseline Year 2010)	
	DECI Gap Ages 17–21	DECI Gap Ages 32–36	DECI Gap Ages 17–21	DECI Gap Ages 32–36
1982	0.00	0.00		
1983	0.05	0.02		
1984	0.01	−0.04		
1985	0.01	−0.04		
1986	0.06	0.00		
1987	0.01	−0.02		
1988	0.03	−0.02		
1989	−0.03	−0.07		
1990	−0.02	−0.06		
1991	0.00	−0.06		
1992	0.04	−0.04		
1993	0.07	−0.04		
1994	0.07	−0.02		
1995	0.06	−0.05		
1996	0.05	−0.02		
1997	0.04	−0.03		
1998	−0.01	−0.06		
1999	−0.06	−0.04		
2000	−0.06	−0.08		
2001	−0.03	−0.05		
2002	0.01	−0.02		
2003	0.05	0.02		
2004	0.08	0.05		
2005	0.10	0.08		
2006	0.08	0.08		
2007	0.09	0.08		
2008	0.10	0.08		
2009	0.17	0.11		
2010	0.17	0.14	0.00	0.00
2011	0.17	0.16	0.00	0.02
2012	0.22	0.16	0.06	0.03
2013	0.20	0.16	0.03	0.03
2014	0.16	0.17	−0.02	0.04
2015	0.15	0.17	−0.02	0.04
2016	0.15	0.16	−0.02	0.03
2017	0.12	0.14	−0.06	0.00
2018	0.08	0.14	−0.10	0.01
2019	0.09	0.12	−0.09	−0.02

Table F.17
Data Points for Figure D.8

Year	Panel A (Baseline Year 1982)		Panel B (Baseline Year 2010)	
	DECI Gap Enlisted	DECI Gap Officers	DECI Gap Enlisted	DECI Gap Officers
1982	0.00	0.00		
1983	0.03	0.01		
1984	−0.01	−0.05		
1985	−0.01	−0.06		
1986	0.04	−0.03		
1987	0.00	−0.07		
1988	0.01	−0.07		
1989	−0.02	−0.17		
1990	−0.01	−0.16		
1991	0.00	−0.17		
1992	0.03	−0.14		
1993	0.05	−0.12		
1994	0.06	−0.14		
1995	0.05	−0.16		
1996	0.05	−0.14		
1997	0.04	−0.15		
1998	0.02	−0.20		
1999	0.01	−0.22		
2000	0.00	−0.26		
2001	0.02	−0.22		
2002	0.04	−0.18		
2003	0.09	−0.12		
2004	0.11	−0.09		
2005	0.14	−0.07		
2006	0.13	−0.06		
2007	0.14	−0.06		
2008	0.13	−0.07		
2009	0.18	−0.05		
2010	0.20	−0.01	0.00	0.00
2011	0.21	0.03	0.02	0.03
2012	0.22	0.02	0.03	0.03
2013	0.23	0.03	0.04	0.04
2014	0.21	0.05	0.01	0.06
2015	0.22	0.04	0.02	0.05
2016	0.20	0.02	0.00	0.02
2017	0.12	0.14	−0.02	0.02
2018	0.08	0.14	−0.03	0.03
2019	0.09	0.12	−0.05	0.00

References

Alhassan, Osman, "Holiday Employment in Retail Trade," *Beyond the Numbers*, Bureau of Labor Statistics, Vol. 8, No. 13, October 2019. As of April 1, 2020:
https://www.bls.gov/opub/btn/volume-8/holiday-employment-in-retail-trade.htm

Asch, Beth J., *Navigating Current and Emerging Army Recruiting Challenges*, Santa Monica, Calif.: RAND Corporation, RR-3107-A, 2019a. As of July 1, 2020:
https://www.rand.org/pubs/research_reports/RR3107.html

Asch, Beth J., *Setting Military Compensation to Support Recruitment, Retention, and Performance*, Santa Monica, Calif.: RAND Corporation, RR-3197-A, 2019b. As of July 1, 2020:
https://www.rand.org/pubs/research_reports/RR3197.html

Asch, Beth J., Paul Heaton, James Hosek, Paco Martorell, Curtis Simon, and John T. Warner, *Cash Incentives and Military Enlistment, Attrition, and Reenlistment*, Santa Monica, Calif.: RAND Corporation, MG-950-OSD, 2010. As of October 2, 2018:
https://www.rand.org/pubs/monographs/MG950.html

Asch, Beth J., James Hosek, and Craig Martin, *A Look at Cash Compensation for Active-Duty Military Personnel*, Santa Monica, Calif.: RAND Corporation, MR-1492-OSD, 2002. As of February 1, 2019:
https://www.rand.org/pubs/monograph_reports/MR1492.html

Asch, Beth J., James Hosek, and John T. Warner, *An Analysis of Pay for Enlisted Personnel*, Santa Monica, Calif.: RAND Corporation, DB-344-OSD, 2001. As of August 7, 2020:
https://www.rand.org/pubs/documented_briefings/DB344.html

Asch, Beth J., John A. Romley, and Mark E. Totten, *The Quality of Personnel in the Enlisted Ranks*, Santa Monica, Calif.: RAND Corporation, MG-324-OSD, 2005. As of October 2, 2018:
https://www.rand.org/pubs/monographs/MG324.html

Asch, Beth J., and John T. Warner, "Recruiting and Retention to Sustain a Volunteer Military Force," in David Galbreath and John Deni, eds. *Routledge Handbook of Defence Studies*, Taylor and Francis Group, 2018, Chapter 7. As of May 6, 2020:
https://www.routledge.com/Routledge-Handbook-of-Defence-Studies/Galbreath-Deni/p/book/9781138122505#toc

Blau, Francine D., and Lawrence M. Kahn, "The Gender Wage Gap: Extent, Trends, and Explanations," *Journal of Economic Literature*, Vol. 55, No. 3, 2017, pp. 789–865.

BLS—*See* Bureau of Labor Statistics.

Buddin, Richard, *Analysis of Early Military Attrition Behavior*, Santa Monica, Calif.: RAND Corporation, R-3069-MIL, 1984. As of August 17, 2020:
https://www.rand.org/pubs/reports/R3069.html

Buddin, *Trends in Attrition of High-Quality Military Recruits*, Santa Monica, Calif.: RAND Corporation, R-3539-FMP, 1988. As of August 17, 2020:
https://www.rand.org/pubs/reports/R3539.html

Bureau of Labor Statistics, National Compensation Survey, *Employment Cost Index Historical Listing—Volume V*, "Continuous Occupational and Industry Series September 1975–March 2020 (December 2005 = 100)," April 2020. As of July 8, 2020:
https://www.bls.gov/web/eci/ecicois.pdf

Clark, Kelly A., "The Job Openings and Labor Turnover Survey: What Initial Data Show," *Monthly Labor Review*, November 2004, pp. 14–23. As of April 1, 2020:
https://www.bls.gov/opub/mlr/2004/11/art2full.pdf

Congressional Research Service, *Defense Primer: Military Pay Raise*, January 6, 2020.

Cortes, Patricia, and Jessica Pan, "Occupation and Gender," Institute of Labor Economics, Discussion Paper No. 10672, 2017.

Couch, Kenneth A., and Dana W. Placzek, "Earnings Losses of Displaced Workers Revisited," *American Economic Review*, Vol. 100, No. 1, 2010, pp. 572–589.

CPS—*See* U.S. Census Bureau and U.S. Bureau of Labor Statistics, *Annual Social and Economic Supplement (ASEC) of the Current Population Survey (CPS)*, 1980–2019.

Danielson, Melissa L., Rebecca H. Bitsko, Reem M. Ghandour, Joseph R. Holbrook, Michael D. Kogan, and Stephen J. Blumberg, "Prevalence of Parent-Reported ADHD Diagnosis and Associated Treatment Among U.S. Children and Adolescents, 2016," *Journal of Clinical Child and Adolescent Psychology*, Vol. 47, No. 2, 2018, pp. 199–212.

Davis, Steven J., and Till M. von Wachter, "Recessions and the Cost of Job Loss," National Bureau of Economic Research, Working Paper No. 17638, 2017. As of April 24, 2020:
https://www.nber.org/papers/w17638

Dertouzos, James N., *Recruiter Incentives and Enlistment Supply*, Santa Monica, Calif.: RAND, R-3065-MIL, 1985. As of August 17, 2020:
https://www.rand.org/pubs/reports/R3065.html

Dertouzos, James N., and Steven Garber, *Human Resource Management and Army Recruiting: Analyses of Policy Options*, Santa Monica, Calif.: RAND Corporation, MG-433-A, 2006. As of October 4, 2018:
https://www.rand.org/pubs/monographs/MG433.html

DoD—*See* U.S. Department of Defense.

Ewing, Philip, "The Military-Civilian 'Disconnect,'" Politico.com, February 20, 2011. As of April 23, 2020:
https://www.politico.com/story/2011/02/the-military-civilian-disconnect-049838

Flood, Sarah, Miriam King, Renae Rodgers, Steven Ruggles, and J. Robert Warren, Integrated Public Use Microdata Series, Current Population Survey: Version 7.0 [dataset], Minneapolis, Minn.: IPUMS, 2020. As of July 8, 2020:
https://doi.org/10.18128/D030.V7.0

Ford, M., B. Griepentrog, K. Helland, and S. Marsh, *The Strength and Viability of the Military Propensity-Enlistment Relationship: Evidence from 1995–2003*, Washington, D.C.: Joint Advertising and Marketing Research Studies, Office of the Undersecretary of Defense for Personnel and Readiness, U.S. Department of Defense, JAMRS Report No. 2009-005, April 2009, not available to the general public.

Fricker, Ronald D., James Hosek, and Mark E. Totten, *How Does Deployment Affect Retention of Military Personnel?* Santa Monica, Calif.: RAND Corporation, RB-7557-OSD, 2003. As of April 2, 2020:
https://www.rand.org/pubs/research_briefs/RB7557.html

Goldberg, Lawrence, *Enlisted Supply: Past, Present, and Future*, Alexandria Va.: Center for Naval Analyses, CNS 1168-Vol. 1, 1982.

Goldich, Robert L., *Military Pay and Benefits: Key Questions and Answers*, Washington, D.C.: Congressional Research Service, June 2005. As of July 1, 2020:
https://fas.org/sgp/crs/natsec/IB10089.pdf

Green, Bert F., and Anne S. Mavor, eds., *Modeling Cost and Performance for Military Enlistment: Report of a Workshop*, Committee on Military Enlistment Standards, Commission on Behavioral and Social Sciences and Education, National Research Council, Washington, D.C.: National Academy Press, 1994.

Greenbooks—*See* Office of the Under Secretary of Defense for Personnel and Readiness, Directorate of Compensation, *Selected Military Compensation Tables*, 1980–2018.

Hardison, Chaitra M., Tracy C. McClausland, Michael G. Shanley, Anna Rosefsky Saavedra, Jaclyn Martin, Jonathan P. Wong, Angela Clague, and James C. Crowley, *Methodology for Translating Enlisted Veterans' Nontechnical Skills into Civilian Employers' Terms,* Santa Monica, Calif.: RAND Corporation, RR-1919-OSD, 2017. As of July 1, 2020:
https://www.rand.org/pubs/research_reports/RR1919.html

Hogan, Paul, Javier Espinosa, Patrick Mackin and Peter Greenston, *A Model of Reenlistment Behavior: Estimates of the Effects of Army's Selective Reenlistment Bonus on Retention by Occupation,* Arlington, Va.: U.S. Army Research Institute for the Behavioral and Social Sciences, June 2005. As of June 2020:
http://citeseerx.ist.psu.edu/viewdoc/download?doi=10.1.1.515.7370&rep=rep1&type=pdf

Hosek 1992—*See* Hosek, James, Christine E. Petersen, Jeannette Van Winkle, and Hui Wang, *A Civilian Wage Index for Defense Manpower,* Santa Monica, Calif.: RAND Corporation, R-4190-FMP, 1992.

Hosek, James, Beth J. Asch, and Michael G. Mattock, *Should the Increase in Military Pay Be Slowed?* Santa Monica, Calif.: RAND Corporation, TR-1185-OSD, 2012. As of July 1, 2020:
https://www.rand.org/pubs/technical_reports/TR1185.html

Hosek, James, Beth J. Asch, Michael G. Mattock, and Troy D. Smith, *Military and Civilian Pay Levels, Trends, and Recruit Quality,* Santa Monica, Calif.: RAND Corporation, RR-2396-OSD, 2018. As of February 1, 2019:
https://www.rand.org/pubs/research_reports/RR2396.html

Hosek, James, Christine E. Peterson, and Joanna Zorn Heilbrunn, *Military Pay Gaps and Caps,* Santa Monica, Calif.: RAND Corporation, MR-368-P&R, 1994. As of April 27, 2020:
https://www.rand.org/pubs/monograph_reports/MR368.html

Hosek, James, Christine E. Petersen, Jeannette Van Winkle, and Hui Wang, *A Civilian Wage Index for Defense Manpower,* Santa Monica, Calif.: RAND Corporation, R-4190-FMP, 1992. As of January 31, 2019:
https://www.rand.org/pubs/reports/R4190.html

Hosek, James, and Jennifer Sharp, *Keeping Military Pay Competitive: The Outlook for Civilian Wage Growth and Its Consequences,* Santa Monica, Calif.: RAND Corporation, IP-205-A, 2001. As of July 1, 2020:
https://www.rand.org/pubs/issue_papers/IP205.html

Hosek, James, and Mark E. Totten, *Serving Away from Home: How Deployments Influence Reenlistment,* Santa Monica, Calif.: RAND Corporation, MR-1594-OSD, 2002. As of April 2, 2020:
https://www.rand.org/pubs/monograph_reports/MR1594.html

IPUMS—*See* Flood, Sarah, Miriam King, Renae Rodgers, Steven Ruggles, and J. Robert Warren, Integrated Public Use Microdata Series, Current Population Survey: Version 7.0 [dataset], Minneapolis, Minn.: IPUMS, 2020.

Jacobson, Louis S., Robert J. LaLonde, and Daniel G. Sullivan, "Earnings Losses of Displaced Workers," *American Economic Review,* Vol. 83, No. 4, 1993.

Keyes, Katherine M., Dahsan Gary, Patrick M. O'Malley, Ava Hamilton, and John Schulenberg, "Recent Increases in Depressive Symptoms Among US Adolescents: Trends from 1991 to 2018," *Social Psychiatry and Psychiatric Epidemiology,* Vol. 54, No. 8, August 2019, pp. 987–996.

Kilmer, Beau and Robert J. MacCoun, "How Medical Marijuana Smoothed the Transition to Marijuana Legalization in the United States," *Annual Review of Law and Social Science,* Vol. 13, 2017, pp. 181–202.

Knapp, David, Bruce R. Orvis, Christopher E. Maerzluft, and Tiffany Tsai, *Resources Required to Meet the U.S. Army's Enlisted Recruiting Requirements Under Alternative Recruiting Goals, Conditions, and Eligibility Policies,* Santa Monica, Calif.: RAND Corporation, RR-2364-A, 2018. As of April 23, 2020:
https://www.rand.org/pubs/research_reports/RR2364.html

Lin, Dajun, Randall Lutter, and Christopher Ruhm, "Cognitive Performance and Labour Market Outcomes, *Labour Economics,* Vol. 51, 2018, pp. 121–135.

Mayberry, Paul, and Neil Carey, "The Effect of Aptitude and Experience on Mechanical Job Performance," *Educational and Psychological Measurement,* Vol. 57, No. 1, 1997, pp. 131–149.

National Center for Education Statistics (NCES), Institute of Education Sciences, "Number and Percentage Distribution of First-Time Postsecondary Students Starting at 2- and 4-Year Institutions During the 2011–12 Academic Year, by Attainment and Enrollment Status and Selected Characteristics: Spring 2014," *Digest of Education Statistics*, Table 326.50, February 2017. As of July 16, 2019: https://nces.ed.gov/programs/digest/d17/tables/dt17_326.50.asp

National Commission on Military, National, and Public Service, *Inspired to Serve: The Final Report of the National Commission on Military, National, and Public Service*, March 2020. As of July 8, 2020: https://inspire2serve.gov/sites/default/files/final-report/Final%20Report.pdf

Office of the Under Secretary of Defense for Personnel and Readiness, *Military Compensation Background Papers: Compensation Elements and Related Manpower Cost Items, Their Purposes and Legislative Backgrounds, Eighth Edition*, Washington, D.C.: U.S. Department of Defense, July 2018. As of May 6, 2020: https://militarypay.defense.gov/Portals/3/Documents/Reports/Mil-Comp_8thEdition.pdf?ver=2018-09-01-181142-307

Office of the Under Secretary of Defense for Personnel and Readiness, *Population Representation in the Military Services: Fiscal Year 2018, Appendix D: Historical Data Tables, Table D-2: Annual Civilian Unemployment Rate by Age Group, 1973–2018*, 2020. As of April 8, 2020: https://www.cna.org/research/pop-rep

Office of the Under Secretary of Defense for Personnel and Readiness, Directorate of Compensation, *Selected Military Compensation Tables*, 1975–2018. As of July 6, 2020: https://militarypay.defense.gov/References/Greenbooks/

OUSD(P&R)—*See* Office of the Under Secretary of Defense for Personnel and Readiness.

Orvis, Bruce R., Michael T. Childress, and Michael Polich, *The Effect of Personnel Quality on the Performance of Patriot Air Defense System Operators*, Santa Monica, Calif.: RAND Corporation, R-3901-A, 1992. As of July 1, 2020: https://www.rand.org/pubs/reports/R3901.html

Orvis, Bruce R., Christopher E. Maerzluft, Sung-Bou Kim, Michael G. Shanley, and Heather Krull, *Prospective Outcome Assessment for Alternative Recruit Selection Policies*, Santa Monica, Calif.: RAND Corporation, RR-2267-A, 2018. As of August 5, 2019: https://www.rand.org/pubs/research_reports/RR2267.html

Pacula, Rosalie L., and Rosanna Smart, "Medical Marijuana and Marijuana Legalization," *Annual Review of Clinical Psychology*, Vol. 13, May 8, 2017, pp. 397–419.

Parker, Kim, Ruth Igielnik, Amanda Barroso, and Anthony Cilluffo, "The American Veteran Experience and the Post-9/11 Generation: 3. The Transition to Post-Military Employment," Pew Research Center, 2019. As of April 2, 2020: https://www.pewsocialtrends.org/2019/09/09/the-transition-to-post-military-employment/

Pierce, Justin R., and Peter K. Schott, "The Surprisingly Swift Decline of US Manufacturing Employment," *American Economic Review*, Vol. 106, No. 7, 2016, pp. 1632–1662.

Pop Rep—*See* Office of the Under Secretary of Defense for Personnel and Readiness, *Population Representation in the Military Services: Fiscal Year 2018, Appendix D: Historical Data Tables, Table D-2: Annual Civilian Unemployment Rate by Age Group, 1973–2018*, 2020.

"Project A: The U.S. Army Selection and Classification Project (Special Issue)," *Personnel Psychology*, Vol. 43, No. 2, 1990.

Public Law 101-509, Treasury, Postal Service and General Government Appropriations Act, Section 529: Federal Employees Pay Comparability Act of 1990, November 5, 1990.

Public Law 108-136, National Defense Authorization Act for Fiscal Year 2004, November 24, 2003.

Ramsberger, Peter, Janice Laurence, Rodney McCloy, and Ani DiFazio, *Augmented Selection Criteria for Enlisted Personnel*, Alexandria, Va.: Human Resources Research Organization, 1999. As of June 10, 2020: https://apps.dtic.mil/sti/citations/ADA363068

Rivera Drew, Julia A., Sarah Flood, and John Robert Warren, "Making Full Use of the Longitudinal Design of the Current Population Survey: Methods for Linking Records Across 16 Months," *Journal of Economic and Social Measurement*, Vol. 39, No. 3, 2014, pp. 121–144.

Ruser, John W., "The Employment Cost Index: What Is It?" *Monthly Labor Review*, Vol. 124, No. 9, 2001, pp. 3–16.

Sasser Modestino, Alicia, Daniel Shoag, and Joshua Ballance, "Downskilling: Changes in Employer Skill Requirements over the Business Cycle," *Labour Economics*, Vol. 41, 2016, pp. 333–347.

Sasser Modestino, Alicia, Daniel Shoag, and Joshua Ballance, "Upskilling: Do Employers Demand Greater Skills When Workers Are Plentiful?" *Review of Economics and Statistics,* posted online June 4, 2019. As of April 2, 2020:
https://www.mitpressjournals.org/doi/abs/10.1162/rest_a_00835?mobileUi=0

Sellman, W. Steven, *Public Policy Implications for Military Entrance Standards*, Keynote address presented at the 39th Annual Conference of the International Military Testing Association, Sydney, Australia, 1997.

Sellman, W. Steven, "Predicting Readiness for Military Service. How Enlistment Standards are Established," manuscript prepared for the National Assessment Government Board, September 30, 2004.

Simon, Curtis J., and John T. Warner, "Managing the All-Volunteer Force in a Time of War," *Economics of Peace and Security Journal*, Vol. 2, No. 1, 2007, pp. 20–29.

Smith, Troy D., Beth J. Asch, and Michael G. Mattock, *An Updated Look at Military and Civilian Pay Levels and Recruit Quality*, Santa Monica, Calif.: RAND Corporation, RR-3254-OSD, 2020.

Smith, D. Alton, and Paul Hogan, "The Accession Quality Cost/Performance Trade-Off Model," in Bert F. Green and Anne S. Mavor, eds., in *Modeling Cost and Performance for Military Enlistment*, Washington, D.C.: National Academy Press, 1994, pp. 105–128.

SOFS—*See* U.S. Department of Defense, Office of People Analytics, Status of Forces Survey, 2012–2018.

U.S. Census Bureau, *Current Population Survey Design and Methodology*, Technical Paper 66, 2006. As of April 1, 2020:
https://www.census.gov/prod/2006pubs/tp-66.pdf

U.S. Census Bureau and U.S. Bureau of Labor Statistics, *Annual Social and Economic Supplement (ASEC) of the Current Population Survey (CPS)*, 1980–2019. As of July 8, 2020:
https://www.census.gov/programs-surveys/saipe/guidance/model-input-data/cpsasec.html

U.S. Census Bureau and U.S. Bureau of Labor Statistics, "CPI for All Urban Consumers (CPI-U): All Items in U.S. City Average, All Urban Consumers, Not Seasonally Adjusted," Series Id:CUUR0000SA0, 2020. As of July 8, 2020:
https://data.bls.gov/timeseries/cuur0000sa0?series_id=cwur0000s

U.S. Code, Title 37: Pay and Allowances of the Uniformed Services; Chapter 19: Administration; Section 1009: Adjustments of monthly basic pay.

U.S. Department of Defense, *The Report of the President's Commission on an All-Volunteer Force*, Washington, D.C., 1970.

U.S. Department of Defense, *Report of the Seventh Quadrennial Review of Military Compensation*, Washington, D.C., August 21, 1992.

U.S. Department of Defense, *Report of the Ninth Quadrennial Review of Military Compensation*, Washington, D.C., 2002.

U.S. Department of Defense, *Report of the Eleventh Quadrennial Review of Military Compensation: Main Report*, Washington, D.C., 2012.

U.S. Department of Defense, Office of People Analytics, Youth Attitude Tracking Study and Youth Polls, 1984–2018.

U.S. Department of Defense, Office of People Analytics, Status of Forces Survey, 2012–2018.

Warner, John T., "The Influence of Economic Factors on Military Recruiting and Retention: Evidence from Past Studies," unpublished manuscript, Falls Church, Va.: The Lewin Group, October 2010.

Warner, John T., Curtis J. Simon, and Deborah N Payne, *Enlistment Supply in the 1990's: A Study of the Navy College Fund and Other Enlistment Incentive Programs*, Arlington, Va.: Defense Manpower Data Center, DMDC Report No. 2000-015, April, 2001. As of October 8, 2018:
http://www.dtic.mil/dtic/tr/fulltext/u2/a390845.pdf

Warner, John T., Curtis J. Simon, and Deborah N Payne, "Propensity, Application, and Enlistment: Evidence from the Youth Attitude Tracking Survey," Clemson University, unpublished manuscript, 2002.

Wenger, Jeffrey, Ellen M. Pint, Tepring Piquado, Michael G. Shanley, Trinidad Beleche, Melissa A. Bradley, Jonathan Welch, Laura Werber, Cate Yoon, Eric J. Duckworth, and Nicole H. Curtis, *Helping Soldiers Leverage Army Knowledge, Skills, and Abilities in Civilian Jobs*, Santa Monica, Calif.: RAND Corporation, RR-1719, 2017. As of July 1, 2020:
https://www.rand.org/pubs/research_reports/RR1719.html

Wenger, Jennie W., Zachary T. Miller, and Seema Sayala, *Recruiting in the 21st Century: Technical Aptitude and the Navy's Requirements*, Washington, D.C.: CNA Analysis and Solutions, CRM D0022305.A2/Final, May 2010. As of April 23, 2020:
https://www.cna.org/CNA_files/PDF/D0022305.A2.pdf